Liquid Crystals
and
Biological Structures

Liquid Crystals
and
Biological Structures

Glenn H. Brown
Liquid Crystal Institute
Kent State University
Kent, Ohio

Jerome J. Wolken
Carnegie–Mellon University
Mellon Institute of Science
Pittsburgh, Pennsylvania

Academic Press   New York   San Francisco   London   1979

A Subsidiary of Harcourt Brace Jovanovich, Publishers

ACADEMIC PRESS, INC.
111 Fifth Avenue, New York, New York 10003

United Kingdom Edition published by
ACADEMIC PRESS, INC. (LONDON) LTD.
24/28 Oval Road, London NW1 7DX

Library of Congress Cataloging in Publication Data

Brown, Glenn Halstead.
   Liquid crystals and biological structures.

   Includes bibliographies.
   1.   Molecular biology.   2.   Liquid crystals.
3.   Cell receptors.   I.   Wolken, Jerome J., joint author.
II.   Title.
QH506.B765   574.1'92   78–67873
ISBN 0–12–136850–5

PRINTED IN THE UNITED STATES OF AMERICA

79 80 81 82   9 8 7 6 5 4 3 2 1

Contents

RECEPTORS, EFFECTORS, AND MEMBRANES

Preface

A century before chemists and physicists began investigating the liquid crystalline state, biologists recognized that the living cell possessed properties that we now call liquid crystalline. It was during the Faraday Society Meeting in 1933 that the liquid crystalline nature of biological structures emerged. By the time of the First International Conference in Liquid Crystals at Kent State University in 1965, liquid crystals and biological structures became an important subject for discussion. In the 1970s we started a dialogue to explore to what extent the properties and structures of liquid crystals are associated with biological structures and to life processes.

In this book we have tried to indicate some of the relationships between liquid crystals and biological structures. We began with a review of the classification, properties, and structures of liquid crystals. We then selected certain cellular structures that exhibit liquid crystalline behavior. These examples were chosen to emphasize the nature of various cellular membranes, from the cell membrane to receptor and to effector membranes. We did not attempt to cover all aspects of liquid crystals or all biological structures. For the reader seeking further information on liquid crystals, there is an Advances comprised of collected papers dealing with recent research. In addition to specific references given in each chapter, there are several books on the chemistry, physics, and structures of liquid crystals. We have listed these books under the General References. It was and is our hope that the subject will excite students and encourage bioscientists to become more aware of the potential of the liquid crystalline state in exploring a variety of biological phenomena.

Glenn H. Brown
Jerome J. Wolken

Acknowledgments

I (G.B.) wish to express my thanks to graduate students who have made suggestions on the presentation of the properties and structures of liquid crystals. I have also profited from many conversations with my colleagues on various aspects of this subject. I would like to thank Academic Press, New York, for permission to reproduce material from previous publications. Specifically, I would like to thank Professor Alfred Saupe, Dr. Mary Neubert, and Professor H. Sackmann. The author acknowledges permission of the editor of *Pramana* and Dr. Adriaan de Vries. Material taken in part from other sources is acknowledged in the legends to figures or tables.

I (J. J. W.) would like to thank the publishers, Appleton-Century-Crofts, New York; Charles Thomas Publishers, Springfield, Illinois; Van Nostrand Reinhold Company, New York; and Academic Press, New York, for permission to reproduce in whole or in part figures and tables from previous publications. I want to thank Mr. Oliver J. Bashor, Jr., for his technical assistance. I am grateful to all who have been associated with me in the Biophysical Research Laboratory and who have contributed in many ways. I would also like to thank and acknowledge those who kindly sent me photographs which were used for figures. These individuals include Professor Sidney Fox, Dr. K. Mühlethaler, Dr. T. Kuwabara, Dr. Arthur Winfree, and Professor A. Saupe. Data taken in part or adapted from other sources are acknowledged in the legends for figures, in the tables, and in the references. Research grants from the Scaife Family Charitable Trust, Pittsburgh; Kresge Foundation, Troy, Michigan; and the Pennsylvania Lions Eye Research Foundation are gratefully acknowledged.

Liquid Crystals
and
the Cell

Chapter 1

Introduction

The aspect of molecular patterns which seems to have been most underestimated in the consideration of biological phenomenon is that found in liquid crystals.

J. Needham (1950)

What are liquid crystals? and what properties do they have that we can associate with living cellular structures? We have set out to explore these questions. But, before we can do so, it is necessary to define the term *liquid crystals* and to indicate their properties. Then we will see what analogies can be drawn between liquid crystals and living structures. As the result, we will try to show that liquid crystals possess structural and behavioral properties that make them unique model systems to investigate a variety of biological phenomena.

It is of historical interest that Virchow, a biologist, studying the structure of living cells described myelin figures in 1854. We now know that myelin figures in cells (lipid–water systems) are liquid crystalline structures. The discovery of an intermediate state of matter—the "liquid crystalline" state—is credited, though, to Friedrich Reinitzer, an Austrian botanist. He prepared the cholesteryl ester, cholesteryl benzoate, and observed that it had two "melting points" with different properties. At 145.5°C, the solid cholesteryl benzoate structure collapsed to form a turbid liquid (now known to be a liquid crystal) and when further heated to 178.5°C, it became transparent. These observations showed that cholesteryl benzoate has three distinct phases: solid, liquid crystal, and liquid. Reinitzer described his experiment in a paper published in the chemical literature in 1888.

Curious about such behavior of these compounds, Reinitzer wrote to Otto Lehmann, a German physicist, to obtain his opinion on his preparations and to encourage him to investigate the physical properties of these compounds—as noted in the following excerpt:

> I venture to ask you to investigate somewhat closer the physical isomerism of the two enclosed substances. Both substances show such striking and beautiful phenomena that I can hopefully expect that they will also interest you to a high degree. .. . . The substance has two melting points, if it can be expressed in such a manner. At 145.5°C, it melts to a cloudy, but fully liquid melt which at 178.5°C suddenly becomes completely clear. On cooling a violet and blue colour phenomenon appears, which then quickly disappears leaving the substance cloudy but still liquid. On further cooling the violet and blue colouration appears again and immediately afterwards the substance solidifies to a white, crystalline mass. The cloudiness on cooling is caused by the star shaped aggregate. On melting of the solid the cloudiness is caused not by crystals but by a liquid which forms *oily streaks* in the melt.
>
>   [From letter Reinitzer (1888) to Lehmann as republished by Kelker (1973)]

Soon thereafter, Lehmann (1904) made a systematic study of organic compounds and found that many of them exhibited properties similar to cholesteryl benzoate. Each of the compounds behaved both as a liquid in its mechanical properties and as a crystalline solid in its optical properties. He showed that the cloudy intermediate phase appeared to have a crystallike structure *Flüssige Kristalle* and originated the term *liquid crystal*. Lehmann (1922) recognized that such properties exhibited by liquid crystals may have analogies to that of the living state.

Friedel (1922) then pointed out that the name liquid crystal is misleading, because the substances are neither real crystals nor real liquids. Friedel proposed calling their state "mesomorphous." Friedel separated these mesomorphic states into three classes. The first class was named *smectic,* implying a relation to soap. The special feature of this class is that the molecules are arranged side-by-side in a series of layers (except smectic D). The second class was named *nematic.* Nematic materials do not display the same degree of regularity in structure that the smectic phase does. However, they possess similar optical properties and have some degree of molecular order. The third class of liquid crystals was described as *cholesteric* because of its relationship to a large number of derivatives of cholesterol. But Friedel did not believe the cholesteric class to be a separate one and considered it to be nematic.

Later an attempt to change the name to "paracrystals" and to rename the subdivisions was made by Rinne (1933). Nevertheless, an important consequence of his work is that naturally occurring liquid crystals, as he and Bernal (1933, 1951) pointed out, are intimately connected with life processes. Although the terms "mesomorphic states," "paracrystals," and "anisotropic liquids" continue to be used to describe the behavior of these compounds, the term *liquid crystal* has come into common usage. A more detailed account of the history of the development of liquid crystals can be found in the *Transactions of the Faraday Society* (1933) and in Kelker (1973).

It will be helpful, before we discuss in detail the classification, structure, and optics of liquid crystals (Chapters 2–4), to indicate the nomenclature and the basis for classifying the various types of liquid crystal structures.

We now recognize two major classes of liquid crystals: the thermotropic and the lyotropic. The thermotropic liquid crystals are either nematic or smectic. The *nematic* (from the Greek) means thread and describes the threadlike structures that are observed under a microscope. In the nematic structure the molecules maintain a parallel or nearly parallel arrangement to each other. They are mobile in three directions and can rotate about one axis. The *smectic* (also derived from the Greek) means grease or slime. The smectic structure (except smectic D) is stratified, with the molecules arranged in layers; their long axes lie parallel to each other in the layers, approximately normal to the plane of the layers. The molecules can move in two directions in the plane and they can rotate about one axis. Within the layers, the molecules can be arranged either in neat rows or can be randomly distributed.

Included with the nematic liquid crystals is a subclass which is referred to as *cholesteric–nematic* liquid crystals and often called "cholesteric" in the literature. Many of these compounds are derivatives of cholesterol. In the cholesteric–nematic liquid crystals the molecules pack in layers about 2000 Å thick. This is in contrast to the smectic structure where layer thickness is about the length of the molecule or about 20 Å. Although most of the molecules in the cholesteric–nematic state are essentially flat, side chains project upward from the plane of each molecule, with some hydrogen atoms extending below. Thus, the direction of the long axis of the molecule in a chosen layer is slightly displaced from the direction of the axis in adjacent layers and produces a helical structure.

These various liquid crystalline structures are schematically illustrated in Fig. 1.1. It should be noted that in real liquids the molecules are randomly arranged (Fig. 1.1a). In the nematic state, the long axes of the molecules lie essentially parallel (Fig. 1.1b) while in the smectic A structure (Fig. 1.1c) the molecules show two-dimensional order. Within a layer the molecules are randomly distributed, while between layers the molecule arrangement is equally spaced. The molecules in smectic C liquid crystals are packed in equidistant layers, as we find with smectic A liquid crystals. However, molecules in a given layer are tilted in relationship to the plane of the layer (Fig. 1.1d). The tilt angle is sensitive to temperature and the molecular geometry of the molecules. In the cholesteric–nematic state (Fig. 1.1e) the molecules are arranged in each layer like those in the ne-

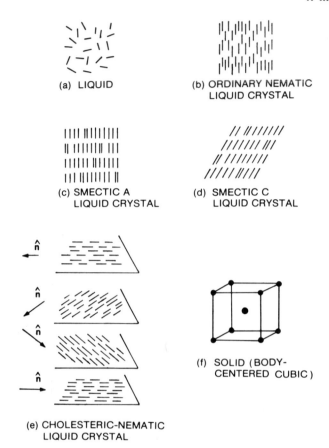

(a) LIQUID

(b) ORDINARY NEMATIC
LIQUID CRYSTAL

(c) SMECTIC A
LIQUID CRYSTAL

(d) SMECTIC C
LIQUID CRYSTAL

(e) CHOLESTERIC–NEMATIC
LIQUID CRYSTAL

(f) SOLID (BODY-
CENTERED CUBIC)

Fig. 1.1 A schematic representation of the (a) molecular arrangement in the isotropic liquid; (b) molecular arrangement in the ordinary nematic liquid crystal; (c) molecular arrangement in the smectic A liquid crystal; (d) molecular arrangement in the
smectic C liquid crystal; (e) molecular arrangement in the cholesteric–nematic liquid
crystal; (f) molecular arrangement in the solid state (a dot represents a molecule).

matic structure, but a necessary twist is superimposed on the layers resulting in a helical structure. The solid crystalline state has order in three
dimensions, as represented by a body-centered cubic packing (Fig. 1.1f).
The liquid crystalline state then has molecular order in one dimension (the
nematic structure) or in two dimensions (for most smectic structures) but
not in three dimensions like a real crystal. Therefore, the liquid crystalline
state is more structured than the liquid state, but less structured than the
solid state.

It should also be noted that liquid crystals are optically anisotropic;

that is, they can transmit light waves of different velocities in different directions. When viewed between crossed Nicol prisms of a polarizing microscope, intense color bands are seen. In the cholesteric–nematic structure, color changes correspond to changes in temperature.

The lyotropic liquid crystals differ from thermotropic liquid crystals, for they are obtained by dispersing a compound with another compound, one of which is a solvent, e.g., water. They are strongly birefringent. Most detergents, soaps, and surfactants dispersed in water will form lyotropic liquid crystals. These compounds are amphiphiles, for they possess in their molecular structure an ionic group that is water soluble and an organic part that is insoluble in water. Lyotropic liquid crystals are biologically important, for they contain two or more components (e.g., lipid–water, lipid–water–protein systems) in which water is an integral part of these liquid crystalline structures.

As we have indicated, an important property of liquid crystals is that they possess both mobility and structural order. In addition, liquid crystals respond to a variety of external stimuli—light, sound, mechanical pressure, temperature, electric and magnetic fields, as well as to changes in its chemical environment—behavioral properties that we associate with living cells.

Bernal (1933) suggested that the living cell is actually a liquid crystal.

A liquid crystal in a cell, through its own structure, becomes a proto-organ for mechanical or electrical activity, and when associated in specialized cells (with others) in higher animals gives rise to true organs, such as muscle and nerve. Secondly, and perhaps more fundamentally, the oriented molecules in liquid crystals furnish an ideal medium for catalytic action, particularly of the complex type needed to account for growth and reproduction. Lastly, a liquid crystal has the possibility of its own structure, singular lines, rods and cones, etc. Such structures belong to the liquid crystal as a unit and not to its molecules, which may be replaced by others without destroying them, and they persist in spite of the complete fluidity of the substance. They are just the properties to be required for a degree of organization between that of the continuous substance, liquid or crystalline solid, and even the simplest living cell.

In search for answers to our question on the relation of liquid crystals to the living cell, we will begin with the structure and properties of liquid crystals (Chapters 2–4). Then we will examine the molecules that comprise cellular structures such as cellular membranes and how they function (Chapters 5–11). With this information, we can begin to develop analogies to the living cell. For example, cellular membranes are structurally lipid–protein–water, that is, liquid crystalline systems and our knowledge of these systems will be applied toward understanding cellular processes on the level of the membrane. Last, we will consider liquid crystals as an analytical and diagnostic tool in medicine (Chapter 12).

This brief introduction to liquid crystals leads us directly to reviewing the structure and behavior of liquid crystals, a necessary requisite before we can discuss living systems.

REFERENCES

Bernal, J. D. (1933). Liquid crystals and anisotropic melts. *Trans. Faraday Soc.* **29**, 1082.

Bernal, J. D. (1951). "The Physical Basis of Life." Routledge and Kegan Paul, London.

Friedel, G. (1922). Les états mésomorphes de la matiére. *Ann. Phys. (Leipzig* [4]) **18**, 273.

Kelker, H. (1973). History of liquid crystals. *Mol. Cryst. Liq. Cryst.* **21**, 1–86.

Lehmann, O. (1904). "Füssige Kristalle, sowie Plästizitat von Kristallen im Allgemeinin, Moledulare Umlagerungen und Aggregatzumstandsänderugen." Englemann, Leipzig.

Lehman, O. (1922). *In* "Handbuch der biologischen Arbeitsmethoden." Physik-Chem. Methoden, Untersuchung des Verhaltens gelöster Stoffe; (E. Arberhalden, ed.) Ab. III, Teil A2, pp. 123–352. Urban and Schwarzenberg, Munich.

Needham, J. (1950). "Biochemistry and Morphogenesis," p. 661. Cambridge Univ. Press, London and New York.

Reinitzer, F. O. (1888). Beitrage zur Kenntniss der Cholesterins. *Monatsh. Chem.* **9**, 421.

Rinne, F. (1933). Investigations and considerations concerning paracrystallinity. *Trans. Faraday Soc.* **29**, 1016.

Transactions of the Faraday Society. (1933). **29**, 881–1085.

Virchow, R. (1854). *Virchows Arch. Pathol. Anat. Physiol.* **6**, 562.

Chapter 2
Classifications of Liquid Crystals

I. INTRODUCTION

There is no simple method to classify the liquid crystalline state of matter. Hosemann *et al.* (1967) have developed a meaningful classification but their classification is too complicated for our consideration in this book.

We have included classifications that are in current usage in liquid crystal sciences. It is recognized that the classifications we present are common usage and are empirical in nature. However, they are not systematic or sophisticated. We have elected to include classifications for both thermotropic and lyotropic liquid crystals since both classifications are found in living systems. Lyotropic systems are the most common in biological structures.

Within a few years a more systematic and sophisticated classification based on structure can be expected to evolve. It should be pointed out that a specific name such as smectic A relates to a molecular packing of a given kind and is therefore useful. The smectic C name also means that a given structure of a given molecular packing exists and that it is structurally different from the smectic A class. However, the names smectic A and smectic C are only empirical and are not a part of a systematic classification procedure.

II. GASEOUS, LIQUID, LIQUID CRYSTALLINE, AND SOLID STATES OF MATTER

Scientists have studied the three states of matter—gases, liquids, and solids—for two centuries. We can consider the liquid crystalline state the fourth state of matter. The liquid crystalline structure exists in both animate and inanimate systems. In life processes their structures are found in cell membranes and in many body tissues. In the laboratory they can be made by heating certain organic molecules (thermotropic liquid crystals) and by mixing two or more components (lyotropic liquid crystals). Certain hydrated metal oxides (e.g., iron, vanadium, and molybdenum) form liquid crystals in aqueous medium.

To give a setting for this book, we shall briefly describe some distinctive characteristics of each of the four states of matter.

A. The Gaseous State

The space in a system constituting a gas is sparsely occupied. Owing to their thermal energy, the molecules of a gas are in continual motion and collide occasionally with one another and with the walls of their container but rebound without loss of energy to the system. Because of the near independence of the particles of a gas, a gas can expand to an unlimited volume. Thus, a distinctive feature of the gaseous state is that its particles (atoms or molecules) are substantially independent of each other except for occasional collisions.

B. The Liquid State

A liquid will take the shape of its container and will bound itself at the top by its own free surface. A primary property of a liquid is that it occupies a certain amount of space; it has a definite density at a given temperature and pressure. The particles of a liquid are close enough to be in contact with one another. In contrast to the gas, the attraction of one molecule to its immediate neighbor is high and, when a liquid is poured, it maintains a constant volume. Furthermore, the molecules in a liquid do not have particular partners as they do in solids, and this irregularity allows a greater degree of tolerance in molecular arrangement. Thus, the liquid state is characterized by its irregularity or indefiniteness in molecular arrangement. If one could use an experimental technique which would be on a small enough observational time scale, the measured properties of a liquid would likely match those of a solid. In other words, the molecules in the liquid state have short-range order. The intermolecular distances are about the size of the molecules.

C. The Solid State

The dominant feature of the crystalline state, in contrast to the gaseous and the liquid states, is the bonding forces between its molecules that give this state its orderly arrangement. Thermal agitation tends to disturb this order in the solid so that, when the temperature is high enough and the average thermal energy of a molecule exceeds its bonding energy, the molecules escape from one another's influence and the solid collapses to a liquid crystal or a liquid. The ability of solids to withstand a shearing force and to regain their original shape after a small deformation are the properties that most readily distinguish them from gases and liquids. Liquid crystals are, to some degree, like solids in their deformation properties. Deformations in liquid crystals are greater than in solids. When a solid is heated, the ordered crystal structure will collapse at a particular temperature, and the resulting phase change is sharp. In characterizing the three common states of matter, one can say that crystalline solids have regular and coherent structure; liquids have irregular and coherent structure; and gases have irregular and incoherent structure.

D. The Liquid Crystalline State

The term *liquid crystals* is at once intriguing and confusing. While it appears self-contradictory, the designation is really an attempt to describe the properties of a particular state of matter. Liquid crystals may be described as condensed fluid states with spontaneous anisotropy.

The liquid crystalline state is a state of matter that mixes the properties of both the liquid and solid states and is intermediate between the two in many of its properties. Liquid crystals combine a kind of long-range order (in the sense of a solid) with the ability to form droplets and to pour (in the sense of waterlike liquids). They also exhibit properties that are found in neither liquids nor solids. Some specific properties include (1) formation of "monocrystals" with application of normal magnetic and/or electric field; (2) optical activity in cholesteric–nematic liquid crystals of a magnitude without parallel in either the solid or liquid regime; and (3) sensitivity (cholesteric–nematic) to temperature, which results in color changes. On heating a solid that forms a liquid crystal, the solid undergoes transformation into a turbid system that is both birefringent and fluid, the consistency varying with different compounds from that of a paste to that of a freely flowing liquid. When the turbid system is heated, it will be converted into the isotropic liquid (optical properties are the same regardless of the direction of measurement). These phase changes can be represented schematically as follows:

$$\boxed{\text{Solid}} \underset{\text{cool}}{\overset{\text{heat}}{\rightleftharpoons}} \boxed{\text{Liquid crystal}} \underset{\text{cool}}{\overset{\text{heat}}{\rightleftharpoons}} \boxed{\text{Liquid}}$$

On cooling the system, the process reverses itself; however, some liquid crystals on cooling will supercool to form an unstable phase (monotropic). Liquid crystals are very responsive to their environment, and external forces have a pronounced effect on their behavior. For example, the application of an electric field of approximately 1.5 V across a cell with a nematic liquid crystal sandwich of 10–25 μm will cause a pronounced change in the optical properties of the liquid crystal.

III. CLASSIFICATION AND NOMENCLATURE OF THERMOTROPIC LIQUID CRYSTALS

Liquid crystals may be divided into two categories, thermotropic and lyotropic. Systems in both categories exhibit polymorphism—that is, more than one liquid crystalline phase can exist for a given compound (thermotropic) or mixture of compounds (lyotropic). The classification of the liquid crystalline structures we know is presented in Fig. 2.1

Some typical organic, thermotropic liquid crystals are recorded in Table 2.1. In this table (Sackmann and Demus, 1966, 1973) we list three different thermotropic nematic liquid crystalline compounds and seven different smectic liquid crystals. Symbols such as S_A mean that we are dealing with a smectic liquid crystal and the subscript A indicates that this is a particular kind of smectic liquid crystal. The subscript A has no significant meaning other than it distinguishes this smectic liquid crystal from six others. We shall consider the two most common nematic liquid crystals (ordinary nematic and cholesteric–nematic) and those smectic liquid crystals most commonly found in the laboratory and in living matter.

A. Nematic Liquid Crystals

1. Ordinary Nematic Structure

Ordinary nematic liquid crystals may be formed by heating organic compounds that are not optically active. Some racemic mixtures will form liquid crystals on heating. The molecules commonly encountered in this class of liquid crystals are elongated, although molecules of other shapes (e.g., disc) are known to show nematic liquid crystallinity. A definitive characteristic of ordinary nematic liquid crystals is the essentially parallel arrangement of the long axes of the molecules. However, the molecular arrangement is such that there is no long-range order in the ordinary nematic liquid crystal.

The arrangement of molecules in two dimensions in an ordinary nematic liquid crystal is represented schematically in Fig. 1.1b. The molecules in this schematic drawing are represented by lines. The molecules are mobile in three directions and can rotate about one axis. This molecu-

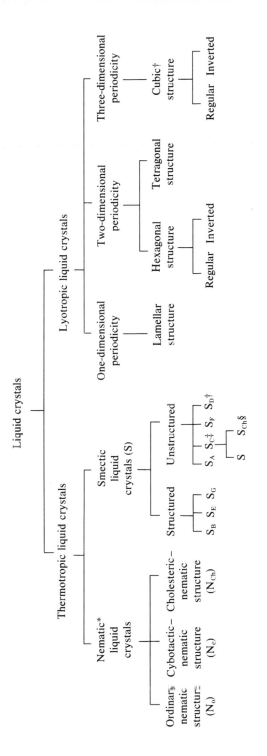

Fig. 2.1. Classification of liquid crystals. * Classes such as skewed-cybotactic nematic liquid crystals and intermediate nematic liquid crystals have been suggested. † Isotropic. ‡ The normal smectic C. § In addition to the normal smectic C, the cholesteric-smectic C structure has been identified. From Brown (1977), *J. Colloid Interface Sci.* **58**, 534. Reprinted with permission.

TABLE 2.1

Some Thermotropic Liquid Crystalline Compounds[a]

Formula	Name	Liquid crystalline range (°C)
I. Nematic liquid crystals		
A. Some ordinary classical nematic liquid crystals		
	p-Methoxybenzylidene-p'-n-butylaniline (MBBA)	21°–47°
	p-Methoxy-p'-n-butylazoxybenzene	19°–76° (mixture of isomers)
	p-Azoxyanisole (PAA)	117°–137°
	p-n-Hexyl-p'-cyanobiphenyl	14°–28°
B. Cholesteric–nematic liquid crystals		
1. Cholesteric esters		
	Cholesteryl nonanoate	145°–179°

2. Noncholesteryl, chiral-type compound

(−)-2-Methylbutyl-p-(p'-methoxybenzylidene-amino)cinnamate   76^c–125°

II. Smectic liquid crystals
A. Structured smectic liquid crystals

Smectic B:

Ethyl p-ethoxybenzal-p'-aminocinnamate   77°–116°

Smectic E:

Diethyl p-terphenyl-p-p''-carboxylate   173°–189°

Smectic G:

2-(p-Pentylphenyl)-5-(p-pentyloxyphenyl)-pyrimidine S_G 79°–103°

Biaxial S_B (or smectic H):

4-Butyloxybenzal-4-ethylaniline   40.5°–51°

(Continued)

TABLE 2.1 (*Continued*)

Formula	Name	Liquid crystalline range (°C)
B. Unstructured smectic liquid crystals		
Smectic A :		
	Ethyl *p*(*p*'-phenylbenzalamino)benzoate	121°–131°
Smectic C :		
n-H$_{17}$C$_8$—O—⟨⟩—COOH	*p*-*n*-Octyloxybenzoic acid	108°–147°
Smectic D :		
	p'-*n*-Octadecyloxy-3'-nitrodiphenyl-*p*-carboxylic acid	159°–195°
Smectic F :		
	2-(*p*-Pentylphenyl)-5-(*p*-pentyloxyphenyl)-pyrimidine	S$_F$ 103°–114°

a From Brown (1977), *J. Colloid Interface Sci.* **58**, 534. Reprinted with permission.

lar arrangement can be compared to a box of round pencils; the pencils can slide and roll back and forth but remain parallel to one another in the direction of their long axes.

2. Cholesteric–Nematic Structure (Commonly Called Cholesteric Liquid Crystals)

Cholesteric–nematic liquid crystals were first observed with cholesteryl esters. Cholesteric–nematic liquid crystals are formed by some optically active organic compounds, or mixtures of the same or by mixing optically active compounds with ordinary nematic liquid crystals. They are miscible with ordinary nematic liquid crystals and have a nematic packing of the molecules in layers with a superimposed twist. This is a helical structure and the distance for a 360° turn is commonly referred to as the pitch.

Twisted nematic liquid crystals (cholesteric–nematic liquid crystals) can be generated by mechanically twisting an ordinary nematic liquid crystal. This can be done easily by rubbing two glass plates in the same direction and then arranging the two plates so that the direction of rubbing of one plate is perpendicular to that of the other. The liquid crystal in this arrangement will rotate polarized light through 90°. Chemical means can also be used to generate cholesteric–nematic structures. These will be discussed later.

B. Smectic Liquid Crystals

Seven smectic structures have been described in the literature. They are identified as smectic A through smectic G.

The term smectic is not particularly definitive and is used for all thermotropic liquid crystals that are not nematics. Six of the seven smectic structures have a lamellar packing of their molecules. The seventh structure gives an isotropic structure. This structure will be discussed later. Smectic liquid crystals may be subdivided into those molecular arrangements which are structured and those which are unstructured.

In most smectic structures, the molecules are arranged in strata; depending on the molecular order within the strata, we can differentiate between smectics with structured and unstructured strata. The structured smectic liquid crystals have long-range order in the arrangement of the molecules in layers and form a regular two-dimensional lattice.

A schematic arrangement in two dimensions for the smectic A structure is sketched in Fig. 1.1c. Lines represent molecules. Notice that the layers are equally spaced and thus give a repeat distance between layers. Within a layer the molecules are randomly arranged. The smectic C structure (Fig. 1.1d) is like the smectic A except the molecules in the layers are tilted with reference to the boundary planes of the layers.

More information on the structure of thermotropic liquid crystals will be presented in Chapter 3.

IV. CLASSIFICATION AND NOMENCLATURE OF LYOTROPIC LIQUID CRYSTALS

A. Introduction

Lyotropic liquid crystals are divided into several categories based primarily on their structure. These different classes are named in Fig. 2.1.

In considering lyotropic liquid crystals we will mention primarily two-component systems composed of water and amphiphilic compounds. However, multicomponent systems are common in lyotropic liquid crystals. An example of such a system which parallels liquid crystals in living systems is lecithin–cholesterol–bile salts–water. Amphiphilic compounds have a polar head (ionic) which tends to dissolve in water (hydrophilic) and a water insoluble organic tail. The molecular geometries found in amphiphilic compounds are of two common types. Type 1 is commonly found in molecules such as sodium stearate. In this type of molecule the polar head is attached to a long hydrophobic tail (water insoluble). Type 2 has the polar head attached to two hydrophobic tails. The hydrophobic groups generally lie side by side and form a "clothespin" structure, or the groups can lie at an acute angle to each other, i.e., the molecules are peg-shaped. Examples of type 2 molecules are Aerosol OT (**I**) and phospholipids.

$$
\begin{array}{c}
\overset{\displaystyle C_2H_5}{|} \\
CH_3-(CH_2)_3-CH-CH_2-OOC-CH_2 \\
CH_3-(CH_2)_3-CH-CH_2-OOC-CH(SO_3^-)Na^+ \\
\underset{\displaystyle C_2H_5}{|}
\end{array}
$$

(I)

Starting with the crystalline form of an amphiphile and water, a series of structures can be generated from the crystal to the true solution. With certain combinations, the polymorphic mesophases formed will show lamellar molecular packing (packing in layers), cubic molecular packing, and hexagonal molecular packing. Removing water can reverse the order of mesophase formation. These statements can be represented schematically as follows:

$$\text{Solid} \underset{-H_2O}{\overset{+H_2O}{\rightleftarrows}} \begin{Bmatrix} \text{Liquid} \\ \text{crystal} \\ \left\{ \begin{matrix} \text{lamellar} \\ \text{structure} \end{matrix} \right\} \end{Bmatrix} \underset{-H_2O}{\overset{+H_2O}{\rightleftarrows}} \begin{Bmatrix} \text{Liquid} \\ \text{crystal} \\ \left\{ \begin{matrix} \text{cubic} \\ \text{structure} \end{matrix} \right\} \end{Bmatrix} \underset{-H_2O}{\overset{+H_2O}{\rightleftarrows}} \begin{Bmatrix} \text{Liquid} \\ \text{crystal} \\ \left\{ \begin{matrix} \text{hexagonal} \\ \text{structure} \end{matrix} \right\} \end{Bmatrix} \underset{-H_2O}{\overset{+H_2O}{\rightleftarrows}}$$

$$\text{Micellar} \underset{-H_2O}{\overset{+H_2O}{\rightleftarrows}} \begin{matrix} \text{Homogeneous} \\ \text{solution} \end{matrix}$$

In the cubic structure the molecules pack in a spherical pattern (see Fig. 3.4b) and the spheres then pack in a cubic design (see Fig. 3.6). In the hexagonal structure the molecules pack in a cylindrical pattern (Fig. 3.4e) and the cylinders or rods pack in a hexagonal structure (see Fig. 3.7). We will now look at these mesophases essentially as they relate to salts of monocarboxylic acids (e.g., sodium stearate) and to certain molecules found in living systems. Schematic representation of molecular packings in lyotropic liquid crystals will be presented in Chapter 3. Let us now look at the different classes of liquid crystals commonly found when an amphiphile and water are mixed.

B. Structures Formed from Amphiphiles and Water

In Fig. 2.1 (Brown, 1977) we have listed the different classes of lyotropic liquid crystals and their relationship with one another. In the paragraphs which follow we give a brief description of each class. More details on structure are given in Chapter 3.

1. The "Neat" or G Phase

It is generally agreed that this phase (Saupe, 1977) is smectic in character, and the amphiphilic molecules with water form a lamellar (layerlike) packing. This will be discussed further in Chapter 3-(see Fig. 3.5 and Table 3.5).

2. The "Middle" or M_1 Phase

This phase is stable at higher water concentrations than the G phase in those cases in which both of these phases are formed from the same components (Saupe, 1977). X-Ray diffraction studies show that the amphiphilic molecules are grouped into rodlike clusters of indefinite length, which, in turn, are arranged side by side in a hexagonal packing (middle phase). It has been proposed that in each rod the molecules are arranged radially around the rod axis with the polar groups on the outside. See Table 3.7 for a description of this phase. A schematic picture of the molecular packing of this structure is presented in Fig. 3.7.

3. The Viscous "Isotropic" or V_1 Phase

This phase appears in some systems at concentrations of amphiphile intermediate between those within which the G and M_1 phases are stable. Ordinary optical observations can give no information on the structure of the phase beyond showing that it is isotropic. X-Ray diffraction studies indicate that the molecules pack in spheres and the spheres then pack in a face-centered cubic lattice or a body-centered cubic lattice (see Table 3.6 and Fig. 3.6 for details on molecular packing of this structure).

4. "Inverse" Phases (V_2 and M_2)

In some systems at concentrations of the amphiphile greater than those at which the G phase is stable, another viscous isotropic phase, V_2, occurs. This phase is followed, with further increase of concentration of amphiphile, by another middle phase, M_2. The M_2 phase has a structure like the M_1 phase, but with the polar groups directed inward and enclosing a water core; the medium between the rods is of hydrocarbon composition.

5. Isotropic S_{1c} Phase

This phase has a higher water concentration than the M_1 phase; the compound n-decyltrimethylammonium chloride, for example, forms the isotropic S_{1c} phase.

If all the mesophases described above occurred for a two-component system at a given temperature, the sequence of their appearance with an increase of amphiphile concentration would be as follows:

$$S_{1c} \longrightarrow M_1 \longrightarrow V_1 \longrightarrow G \longrightarrow V_2 \longrightarrow M_2$$

No system has been found that exhibits all mesophases. The common ones are M_1, V_1, and G, V_2, and M_2. Of these the three most common phases and the order of their existence are as follows:

$$\frac{M_1 \longrightarrow V_1 \longrightarrow G}{\text{Increasing amphiphile concentration}} \longrightarrow$$

The structures of lyotropic phases are discussed in Chapter 3.

REFERENCES

Brown, G. H. (1977). Structure and properties of the liquid crystalline state of matter. *J. Colloid Interface Sci.* **58,** 534.

Friedel, G. (1922). The mesomorphic state of matter. *Ann. Phys., (Leipzig)* [4] **18,** 273.

Hosemann, R., Lemm, K., and Wilke, W. (1967). The paracrystal as a model for liquid crystals. *Mol. Cryst. Liq. Cryst.* **2,** 333.

Sackmann, H., and Demus, D. (1966). The polymorphism of liquid crystals. *Mol. Cryst. Liq, Cryst.* **2,** 81.

Sackmann, H., and Demus, D. (1973). The problems of polymorphism in liquid crystals. *Mol. Cryst. Liq, Cryst.* **21,** 239.

Saupe, A. (1977). Textures, deformations, and structural order of liquid crystals. *J. Colloid Interface Sci.* **58,** 549.

Chapter 3
Structure of Liquid Crystals

I. INTRODUCTION

We learned in Chapter 2 that liquid crystals are divided into two categories. Thermotropic liquid crystals, as the name implies, are formed by heating certain solids; lyotropic liquid crystals are prepared by mixing two or more components. Both categories of liquid crystals exhibit polymorphism; that is, more than one liquid crystalline structure can exist for a given compound (thermotropic) or a mixture of compounds (lyotropic). Examples of phase changes in thermotropic liquid crystals are represented schematically in Fig. 3.1. It should be pointed out that more complex phase relationships are known. These thermotropic liquid crystals, as well as lyotropic phases, will be considered in this chapter. When we look at polymorphism later in this chapter, we will see more clearly that there can be many different liquid crystalline structures. For excellent discussions of liquid crystals see de Gennes (1974) and Chandrasekhar (1977).

Many organic compounds that form liquid crystals when heated have elongated molecules. The axial ratio commonly encountered in many organic molecules that form liquid crystals is 4–8 : 1 (assuming the molecule to be a cylinder, the length is 4 to 8 times its diameter). Molecular weights generally fall in the region of 200 to 500 atomic weight units. It should be remembered that molecules that show liquid crystallinity can have different geometric shapes, e.g., elongated, disc-shaped and molecules with condensed ring systems as found in coal tar.

Lyotropic liquid crystals always require the participation of a "solvent." An example is water and an amphiphile. An amphiphilic compound has in the same molecule two groups which show quite different solubility properties. The hydrophilic portion of the molecule tends to be water soluble and insoluble in organic solvents; the lipophilic portion is water insoluble and soluble in organic solvents. Depending on the relative

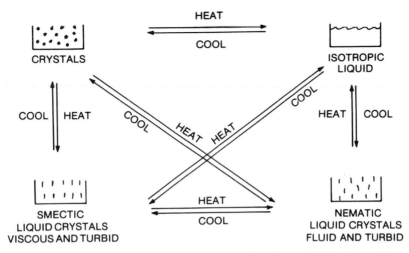

Fig. 3.1 Some phase changes involving nematic and smectic liquid crystals.

contribution of each of the molecular portions, amphiphilic compounds may range from essentially hydrophilic to predominantly lipophilic. The amphiphilic molecules which have the greatest tendency to form liquid crystalline systems with water are those in which the hydrophilic and lipophilic units are strong and rather equally matched.

Although the molecules forming lyotropic systems are large compared to the thermotropic liquid crystalline compounds, their axial ratios are seldom larger than approximately 15. Deoxyribonucleic acid (DNA), certain viruses (e.g., tobacco mosaic), and many synthetic polypeptides form lyotropic liquid crystals with an appropriate second component (usually water) in suitable concentration. Some of these systems are temperature sensitive and they decompose on heating. Many systems are of the kind just cited and will show different polymorphic forms as the concentration and/or temperature changes. The solute molecules often exhibit a solute–solute interaction which produces long-range order. In some of the lyotropic systems the solute–solvent interaction is important. For example, in an amphiphilic–water system, water is an integral part of the structure.

Lyotropic mesomorphous systems are usually as sensitive to changes in temperature as thermotropic systems.

II. MOLECULAR STRUCTURE AND POLYMORPHISM OF THERMOTROPIC LIQUID CRYSTALS

Thermotropic liquid crystals are classified as nematic and smectic; these classes can be further subdivided. In the pages which follow we shall discuss the structure of each class and consider polymorphism exhibited by change in temperature.

A. Ordinary Nematic Structure

The arrangement of molecules in two dimensions in the ordinary nematic liquid crystal is represented schematically in Fig. 3.2. Two features of ordinary nematic liquid crystals are (1) there is a long-range orientational order, i.e., the long axes of the molecules are essentially parallel; (2) the nematic structure is fluid, i.e., there is no long-range correlation of the molecular center of mass positions.

The direction of the principal axis $\hat{n}$ (the director) is arbitrary in space. The only structural restriction in the ordinary nematic liquid crystal is that the long axes of the molecules maintain a parallel or nearly parallel ar-

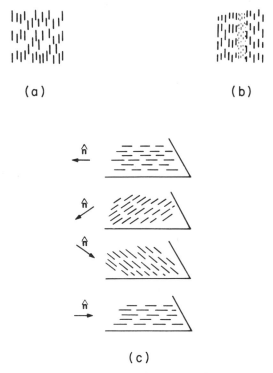

(a)   (b)

(c)

Fig. 3.2 Two-dimensional sketch of molecular packing in (a) ordinary and (b) twisted nematic liquid crystals. (c) Three-dimensional sketch of molecular packing in a cholesteric–nematic liquid crystal. The cholesteric director $\hat{n}$ follows the form of a helix.

rangement. The molecules are mobile in three directions and can rotate about one axis (Fig. 3.2a).

Bulk samples of nematic liquid crystals are turbid. In films greater than 0.1 mm thick, they show threadlike disclinations between crossed polarizers. In thinner films, a schlieren texture with pointlike singularities can be obtained (Figure 4.4). These singularities are vertical threads and may be characterized by the number of dark brushes that appear when observed between crossed polarizers. Points with two or four brushes are commonly found. By simultaneous rotation of the polarizer and analyzer, positive and negative points can be distinguished, depending on whether the brushes rotate in the same (positive) or opposite sense (negative).

Molecules in a film of a nematic liquid can be oriented by surface action. If the surface is rubbed, the molecules tend to align with their long axes parallel to the direction of rubbing. Other treatments of the surface, such as with certain surfactants, may orient the molecules so that their long axes stand perpendicular or parallel to the surface. If the orientation of the long axes of the molecules perpendicular to the surface is complete, a pseudoisotropic texture results. The pseudoisotropic texture appears black when observed between crossed polarizers. If one touches the cover glass flashes of light will appear, indicating that the perpendicular orientation has been destroyed.

The nematic phase is the highest temperature mesophase in thermotropic liquid crystals. It is transformed on heating to the isotropic liquid. This transition is first order; the enthalpy of the transition generally lies between 0.1 and 1.0 kcal/mole. The energy required to deform a liquid crystal is so small that even the slightest perturbation caused by a dust particle or an inhomogeneity on the glass surface on which the liquid crystal rests can distort the structure considerably.

Nematic liquids have an infinite-fold symmetry axis and are, therefore, uniaxial. The orientation of the molecules in a nematic liquid crystal is incomplete. The measure of the degree of orientation can be reasonably expressed by a single order parameter S,

$$S = \tfrac{1}{2}(3 \cos^2\theta - 1)$$

where θ denotes the angle between the long molecular axis and the nematic symmetry axis. Experimental values of S (order parameter) range from 0.4 near the nematic–isotropic point to 0.8 near the nematic–smectic point or nematic–crystal point if no smectic phase exists.

B. Cholesteric–Nematic Structure

The cholesteric–nematic structure (also called the twisted nematic structure) was first observed with cholesteryl esters (Brown, 1976). If the

constituent molecules are optically active, the mesophase will be a cholesteric–nematic liquid crystal. However, if a racemic mixture of the dextro and levo forms exists, then the mesophase will be an ordinary nematic
liquid crystal. Any ordinary nematic liquid crystal can be converted into a
cholesteric one by adding an optically active compound.

In recent years nonsteroidal molecules (Table 2.1) which are optically
active have been found to show the cholesteric–nematic structure. To
distinguish these compounds from the cholesteric esters, we will call
these chiral nematic liquid crystals.

Figure 3.2b is a two-dimensional sketch of the molecular packing in a
cholesteric–nematic structure. This equilibrium structure is represented
in three dimensions in Fig. 3.2c. By studying Figs. 1.1b, 3.2b, and 3.2c, it
can be seen that one can go from the ordinary nematic to the cholesteric–
nematic structure by a mechanical twist of the ordinary nematic packing.
Figure 3.2c shows that the cholesteric–nematic structure is an ordinary
nematic packing on which a twist has been superimposed.

Lack of long-range translational order imparts fluidity to the cholesteric–nematic phase. On a local scale the ordinary nematic and cholesteric–nematic structures are quite similar. However, on a large scale, the
cholesteric director $\hat{n}$ follows a helical form as illustrated in Fig. 3.2c.
When the spatial period is comparable to the optical wavelength, there
will be a strong Bragg diffraction. If the wavelength of the light is in the
visible region of the spectrum, the cholesteric–nematic liquid crystal will
appear brightly colored. The molecular packing in the twisted nematic
gives a helical form whose pitch is temperature sensitive. If the helix has
infinite pitch, one has a cholesteric–nematic liquid crystal. When the
pitch is zero, the system is an ordinary nematic liquid crystal.

When the pitch is very large compared to the wavelength of the incident
beam, a perpendicular beam is broken up into two linearly polarized
waves with their polarization directions perpendicular and parallel to the
alignment axis. When the pitch is reduced, the incident light is converted
into circularly polarized waves and the material shows optical activity
that is very strong when $\lambda/\sqrt{\varepsilon_1}$ approaches the pitch. λ is the wavelength
of the incident light and ε_1 is one of the principal dielectric constants of the
untwisted material. In the range of $\rho\sqrt{\varepsilon_2} \leq \lambda < \rho\sqrt{\varepsilon_1}$, selective reflection takes place. ρ is the pitch of the helix and ε_1 and ε_2 are the two principal dielectric constants of the untwisted material. In this wavelength re-
gion, one of the circularly polarized components is reflected and the other
is transmitted. This selective reflection gives the material an iridescence
that comes from the periodic structure of the cholesteric–nematic molecular packing. The reflection in the visible region follows Bragg's law. For

normal incidence, only first-order reflections are observed; however, with oblique incidence, higher orders can be observed experimentally.

The most sensitive cholesteric–nematic materials, when observed with monochromatic light, show a visible change of the reflected light with a temperature change of 0.001°C. Natural light can be used for less critical applications with a visible change taking place with a temperature change of 0.01°C.

The cholesteric–nematic liquid crystals function as a diffraction grating for visible light. Compare this property with the diffraction of X radiation by crystals such as sodium chloride.

C. Smectic Structures

The term smectic is not exactly specific, as we now use it, but covers all thermotropic liquid crystals that are not nematics. Seven smectic structures have been described in the literature. They are commonly identified as smectic A through smectic G. A new smectic structure has been designated and has been identified as smectic H. Some argue that the smectic H structure is really a tilted smectic B structure; in our discussion we will consider this to be so.

In most smectic structures, the molecules are arranged in strata; depending on the molecular order within the strata, we can differentiate between smectics with structured and unstructured strata.

The thickness of a smectic layer is of the order of the length of the free molecule or double the length. The interlayer attractions are weak compared with the lateral forces between molecules, and consequently the layers are able to slide over each other rather easily.

The smectic liquid crystal is fluid and at the same time anisotropic because of the ease with which the molecules can slide past one another while still remaining parallel. Smectic liquid crystals are fluid but much more viscous than nematic liquid crystals.

Structured smectic liquid crystals have long-range order in the arrangement of the molecules in layers and form a regular two-dimensional lattice. The most common of the structured smectic liquids is the smectic B. The smectic B structure has two different symmetries, $D_{\infty h}$ and C_{2v}. The first of these has a hexagonal packing, with the molecular axis perpendicular to the layers, and is optically uniaxial. The second smectic B has its molecules tilted in the layers; because of its lower symmetry, it is biaxial. The texture of the structured smectic is a modification of the fan and schlieren textures and of the mosaic texture (see Fig. 4.7). The mosaic texture has optically uniform birefringent areas.

Unstructured smectic liquid crystals have molecules packed in layers, and the molecules in the layers are randomly arranged (liquidlike). One of the common unstructured smectic liquid crystals is the smectic A that has the symmetry type $D_{\infty h}$ and is optically uniaxial. The smectic A structure has the molecules arranged in monomolecular layers with the long axes of the molecules perpendicular to the plane of the layers (Fig. 1.1c). The typical texture of the $D_{\infty h}$ symmetry is focal conic; its modification is the fan texture (Fig. 4.5). Another type of unstructured smectic liquid crystal has C_{2v} symmetry and is optically biaxial. The most common smectic liquid crystal of this type is smectic C, which has layers that are monomolecular. The molecules in the layer are tilted (Fig. 1.1d). The textures of C_{2v} symmetry are of the same kind as those of $D_{\infty h}$ but are often more complicated. The typical C_{2v} textures are broken focal conic and broken fan textures.

We shall comment briefly on the structure of smectic A (S_A), smectic B (S_B), and smectic C (S_C) liquid crystals.

1. Smectic A Structure

The molecules in a smectic A liquid crystal are packed in strata and the molecules in a stratum are randomly arranged. The strata show a repeat distance between centers of gravity of molecules in adjacent strata, and an X-ray pattern of the structure shows a sharp ring characteristic of this packing pattern. The layer thickness is essentially identical to a full molecular length. A second ring on the X-ray pattern at about 10° Bragg angle is diffuse in nature, thus showing that the molecules in a stratum are randomly packed (Fig. 3.3).

A densitometer trace of the X-ray film is located directly under the picture of the film. Comparable data are recorded for smectic B and smectic E. Note that smectic E has many diffraction rings, which indicates that the structure approaches that of a crystal. The smectic B structure is considered below.

2. Smectic C Structure

X-Ray diffraction patterns and microscopic studies support the idea that the smectic C structure has a uniform tilting of the molecular axes with respect to the normal layer. The spacing between strata, as determined by X-ray studies, is considerably less than the molecular length, and the difference between these values indicates molecular tilt. The smectic C structure is optically biaxial, which supports the idea of molecular tilt. The tilt angle is generally temperature dependent. However, not all smectic C structures show a change in tilt angle with temperature.

SMECTIC E   SMECTIC B   SMECTIC A

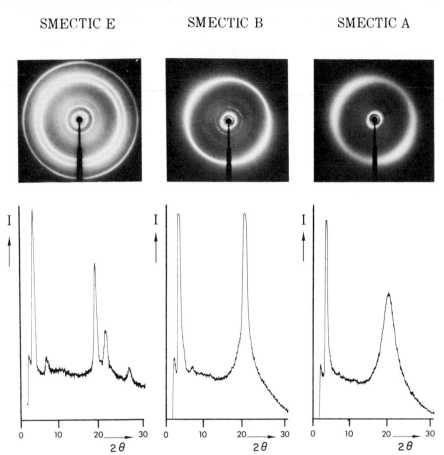

Fig. 3.3 X-Ray diffraction patterns and densitometer traces of these X-ray patterns of S_E, S_B, and S_A (left to right); photographs on same scale. From de Vries (1975), *Pramana Suppl.* **1**, 93. Reproduced by permission.

3. Smectic B Structure

The smectic B structure has the well-ordered layers of molecules and, in addition, orderly packing of the molecules in the layers. X-Ray patterns of a smectic B structure show two sharp rings (see Fig. 3.3), as one would expect from the molecular packing characteristics just described. The smectic B structure has the fluid properties one expects of a liquid crystal, and its mechanical properties are quite different from what one would expect from a material that possesses three-dimensional order.

The smectic B structure has two different symmetries, $D_{\infty h}$ and C_{2v}. The

TABLE 3.1

Polymorphism of Thermotropic Liquid Crystals

Class	Optical properties	Textures	Structure	Examples
I. Nematics				
Ordinary nematic	Uniaxially positive	Schlieren; threaded marbled; pseudo-isotropic; homogeneous	Parallelism of long molecular axes	p-Azoxyanisole; p-methoxybenzylidene p-n-butylaniline
Cholesteric-nematic	Uniaxially negative; or isotropic optically active	Focal conic with Grandjean steps; homogeneous; isotropic	Nematic packing in planes; superimposed twist in direction perpendicular to the long axes of the molecules	Cholesteryl nonanoate
II. Structured Smectics				
Smectic B	Uniaxially or bi-axially positive	Mosaic; stepped drops; pseudo-isotropic; homogeneous; schlieren	Layer structure; molecular axes orthogonal or tilted to the layers; hexagonal arrangement within the layers.	Ethylethoxybenzylidene-aminocinnamate; terephthal-bis-butyl-aniline
Smectic E	Uniaxially positive	Mosaic; pseudo-isotropic	Layer structure; molecular axes orthogonal to the layers; ordered arrangement within the layers	di-n-Propyl-terphenyldi-carboxylate

Smectic G	Uniaxially positive	Mosaic	Layer structure with ordered arrangement within the layers	2-(4-*n*-Pentylphenyl)-5-(4-*n*-pentyloxyphenyl)-pyrimidine
III. Unstructured Smectics				
Smectic A	Uniaxially positive	Focal conic (fan-shaped or polygon); stepped drops; homogeneous; pseudo-isotropic	Layer structure; molecular axes orthogonal to the layers; random arrangement within the layers	Diethylazoxybenzoate
Smectic C	Biaxially positive	Broken focal conic; schlieren; homogeneous	Layer structure; molecular axes tilted to the layers; random arrangement within the layers	Dodecyloxyazoxybenzene
Smectic D	Isotropic	Isotropic; mosaic	Cubic structure	4'-Octadecyloxy-3'-nitro-diphenyl-4-carboxylic acid
Smectic F	Uniaxially positive	Schlieren; broken focal conic with concentric axes	Layer structure	2-(4-*n*-Pentylphenyl)-5-(4-*n*-pentyloxyphenyl)-pyrimidine

first of these has a hexagonal packing, with the molecular axis perpendicular to the layers and is optically uniaxial. The second smectic B has its molecules tilted in the layers; because of its lower symmetry, it is biaxial.

D. General Comments About Polymorphism in Thermotropic Liquid Crystals

Many thermotropic liquid crystals pass through more than one mesophase on heating from the solid to the isotropic phase. Such liquid crystals are said to be "polymorphous." Sackmann and Demus (1966, 1973) have contributed much data on thermotropic polymorphism.

There are two schemes for designating polymorphic forms of smectic liquid crystals. The older of the two was developed by Sackmann and Demus and simply uses the alphabet. The symbols identify the order of discovery of the smectic structure. Therefore, the smectic structures are labeled as S_A through S_G. A classification based on structure established by X ray was proposed by de Vries (1973). de Vries (1979) has just proposed a new classification based on symmetry and order.

TABLE 3.2

Typical Examples of Polymorphic Forms of Thermotropic Liquid Crystals

Polymorphic form[a]	Example
N	 *p*-Azoxyanisole
N	 4-Methoxybenzylidene- 4'-*n*-butylaniline
A	 Diallylazoxybenzene- 4,4'-dicarboxylate
Ch,A	 Cholesteryl nonanoate

TABLE 3.2 (Continued)

Polymorphic form[a]	Example
A,B	Ethyl-4-ethoxybenzylidine-4'-aminocinnamate
N,A,C	4'-n-Hexyloxy-3'-nitro-biphenyl-4-carboxylic acid
A,C,B	n-Amyl-4-n-decyloxybenzyl-idene-4'-aminocinnamate
N,A,C,B	Diethyl terephthalylidene-bis-(4-aminocinnamate)

[a] Key: N = Nematic; Ch = cholesteric; A = Smectic A; B = Smectic B; C = Smectic C.

The order in which different mesophases appear on heating can be easily remembered by utilizing the fact that raising the temperature of a material results in progressive destruction of molecular order. We know that a smectic B is more ordered than a smectic C, a smectic C more than a smectic A, and a smectic A is more ordered than a nematic structure. Therefore, if one has a tetramorphic liquid crystal, the order of stability is

$$\text{Solid} \xrightarrow{\Delta} \text{Smectic B} \xrightarrow{\Delta} \text{Smectic C} \xrightarrow{\Delta} \text{Smectic A} \xrightarrow{\Delta} \text{Nematic} \xrightarrow{\Delta} \text{Isotropic liquid}$$

For substances having nematic and/or smectic structures, but not all of those listed above, the order is identical as illustrated by deleting those structures not present. For example, a trimorphic system involving smectic A, smectic B, and a nematic phase would have the stability

$$\text{Solid} \xrightarrow{\Delta} \text{Smectic B} \xrightarrow{\Delta} \text{Smectic A} \xrightarrow{\Delta} \text{Nematic liquid crystal} \xrightarrow{\Delta} \text{Isotropic liquid}$$

Cholesteric–nematic structures can replace ordinary nematic ones in the scheme.

In Table 3.1 we have summarized the properties and structural characteristics of the different kinds of thermotropic liquid crystals. Table 3.2 contains some typical examples of polymorphic forms of thermotropic liquid crystals.

For further consideration of polymorphism, consider Table 3.3. In the left-hand column we list the degree of polymorphism; in the second column we record the symbols which distinguish monomorphism, dimorphism, trimorphism, and tetramorphism; in the right-hand column we list the kind of phase. M_1 indicates that compound 1 exhibits only one liquid crystalline phase, e.g., nematic; T_E indicates that compound 1 exhibits four mesophases, namely, N, S_A, S_C, and S_B. The collection of compounds in Table 3.3 is not exhaustive but is illustrative.

TABLE 3.3

Degree of Polymorphism Exhibited by Thermotropic Liquid Crystals

Degree of polymorphism[a]	Symbol	Kinds of phases	Degree of polymorphism[a]	Symbol	Kinds of phases
Monomorphism (M)	M_1	N	Trimorphism (T)	T_1	N A B
	M_2	Ch		T_2	N A C
	M_3	A		T_3	Ch A C
	M_4	C		T_4	N C B
				T_5	A B E
				T_6	A C B
Dimorphism (D)	D_1	N A		T_7	A D C
	D_2	Ch A		T_8	A C G
	D_3	N B			
	D_4	N C			
	D_5	Ch C	Tetramorphism (Te)	Te_1	N A C B
	D_6	A B		Te_2	A C F G
	D_7	A C			
	D_8	A E			
	D_9	C B			
	D_{10}	D C			

[a] M, D, T, and Te designate the degree of polymorphism. M designates a monotropic liquid crystal, D designates that the compound D_1 is dimorphic, etc. N = ordinary nematic structure; Ch = cholesteric-nematic structure; A, B, C, D, E, F, and G = smectic structures. Heating is from right to left. For example, D_1 heating and cooling pattern is

$$\text{Isotropic liquid} \underset{\text{heat}}{\overset{\text{cool}}{\rightleftharpoons}} \text{N} \underset{\text{heat}}{\overset{\text{cool}}{\rightleftharpoons}} S_A \underset{\text{heat}}{\overset{\text{cool}}{\rightleftharpoons}} \text{Crystal}$$

III. LYOTROPIC LIQUID CRYSTALS

The term lyotropic liquid crystallinity is used to describe the formation of a thermally stable system by the penetration of a solvent between the molecules of a crystal lattice. Lyotropic liquid crystals require the participation of a solvent. Many of the studies of lyotropic liquid crystals have been observed on lipid systems containing water.

A. Molecular Arrangement

Lyotropic systems are formed by the mixing of two or more components. There are many possible combinations of compounds to form lyotropic liquid crystalline systems. The most common lyotropic liquid crystal systems are amphiphiles and water. Examples are salts of fatty acids in water and phospholipids in water. The general formula for a phospholipid molecule and its solubility characteristics is shown in (**II**). The R and R'

$$
\begin{array}{l}
NH_3^+ \\
| \\
CH_2 \\
| \\
CH_2 \qquad \text{Water-soluble} \\
| \qquad\quad \text{portion of the} \\
O \qquad\quad \text{molecule} \\
| \\
{}^-O-P=O \\
| \\
O \\
----|---- \\
CH_2 \\
| \qquad\qquad \text{Water-insoluble} \\
\;\;\,C\;\; \qquad \text{portion of the} \\
O^{\diagup}\;H^{\diagdown}CH_2 \quad \text{molecule} \\
| \qquad | \\
O=C \quad\; O \\
| \qquad\quad | \\
R' \qquad\; C=O \\
\qquad\qquad | \\
\qquad\qquad R
\end{array}
$$

(II)

chains generally contain 14 to 18 carbon atoms. In the amphiphilic molecule that has a polar head attached to two hydrophilic tails, the organic tails generally lie side by side to one another and form a "clothespin" structure. If the organic tails are at an acute angle to each other, the molecules are arranged in a peg-shape.

Amphiphilic molecules associate in such a pattern that there is a minimum of free energy and the molecular aggregates in both the dry and wet forms are not fundamentally different (Friberg, 1976). In the packing of amphiphilic molecules there are several geometric patterns which are found in nature. On addition of water (Table 3.4) to a crystal composed of amphiphilic molecules the crystal structure collapses to the formation of a lamellar structure. On further addition of water a cubic structure may be

TABLE 3.4

Some Properties of Lyotropic Systems Composed of an Amphiphile and Water

Suggested structural arrangement[a]						
% Water[a] (approximate range)	0	5–22–50	23–40	34–80	30–99.9	Greater than 99.9
Physical state	Crystalline	Liquid crystal-line, lamellar	Liquid crystal-line, face-centered cubic	Liquid crystal-line, hexagonal compact	Micellar solution	Solution
Gross character	Opaque solid	Clear, fluid, moderately viscous	Clear, brittle, very viscous	Clear, viscous	Clear, fluid	Clear, fluid
Freedom of movement	None	2 directions	Possibly none	1 direction	No restrictions	No restrictions
Microscopic properties (crossed nicols)	Birefringent	Neat soap texture	Isotropic with angular bubbles	Middle soap texture	Isotropic with round bubbles	Isotropic
X-Ray data	Ring pattern 3–6 Å	Diffuse halo at about 4.5 Å	Diffuse halo at about 4.5 Å	Diffuse halo at about 4.5 Å		
Structural order	3 dimensions	1 dimension	3 dimensions	2 dimensions	None	None

[a] The different percentages of water show that different amphiphiles require different amounts of water. For soaps, the lamellar struc-ture generally occurs between 5 and 22% water; with some lipophiles the water may be as high as 50%. The cubic structure generally occurs between 23 and 40%.

formed. Further addition of water may give the hexagonal packing, followed by a micellar structure and finally a homogeneous solution. A general description of the structures and their properties will be presented in Chapter 4.

In Fig. 3.4 we show a schematic representation of a number of lyotropic liquid crystalline structures. Figure 3.4a shows a simple monolayer pack-

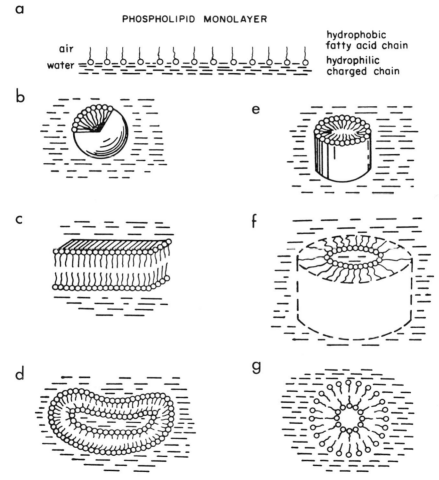

Fig. 3.4 Schematic patterns representing lyotropic mesomorphism of amphiphilic lipids. (a) Monolayer at the air water interface; (b) spherical particle; (c) lamellar phase in water; (d) lamellar phase: section through a sonicated liposome; (e) hexagonal phase I (oil in water); (f) hexagonal phase II (water in oil). (g) cross-section of cylinder with a bilayer packing. The circles represent the polar heads of the lipid molecules.

ing; the white circles represent the ionic head of the molecule while the "tail" is the organic portion of the molecule. Figure 3.4b illustrates a spherically shaped particle. Figure 3.4c represents a lamellar packing (bilayer) while 3.4d sketches a lamellar phase section through a sonicated liposome. Figures 3.4e and 3.4f are sketches of the molecular packing in a regular hexagon and an inverted hexagon. In Fig. 3.4g we have sketched a cross section of a cylinder generated by combining a regular and an inverted molecular packing (bilayer). In all of these figures the circle represents the ionic part of the amphiphilic molecule.

Not all amphiphiles and water will generate all the structures cited in Chapter 2, Section IV, B. It must be remembered that water is an integral part of these structures. The systems are also temperature sensitive, and by changing temperature the equilibria can be shifted.

B. Structure of Lyotropic Liquid Crystals

1. Lamellar Structure

The most common lyotropic liquid crystal has a lamellar structure. In the detergent industry, it is known as the neat soap phase. It corresponds to the smectic A structure found in thermotropic liquid crystals. The molecular packing in the lamellar structure gives double layers with the water-insoluble tails dissolving in each other and the ionic part of the molecule dissolving in water. The double layers pack parallel to one another

LIPOPHILIC GROUP

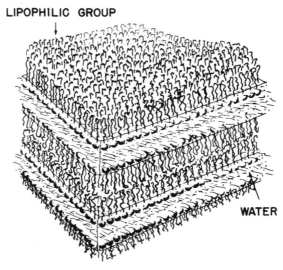

WATER

Fig. 3.5 Schematic representation of lamellar packing of amphiphilic molecules with water. Modified with permission from Rosevear (1968).

TABLE 3.5

Lamellar Packing of Lyotropic Liquid Crystals[a]

Designation	Optical properties	Basic structure	Description of proposed structure
Neat-phase (soap— Boiler's Neat Soap); lamellar	Anisotropic	Lamellar double layers	Double layers of amphiphiles with polar groups in the interfaces with intervening layers of water
Single layered neat phase	Anisotropic	Lamellar single layers	Single layers of amphiphilic molecules oriented with polar groups toward opposite interfaces with intervening layers of water
Mucous woven phase	Slightly anisotropic	Lamellar double layers	Double layers of amphiphiles with polar groups in the interfaces with intervening layers of water.

[a] (1) Structural arrangement displaying Bragg spacing ratio $1:1/2:1/3$. (2) One-dimensional periodicity. (3) Layer structure.

and are separated from each other by a water layer (Fig. 3.5). In other words, the ion heads are anchored in the water layer. This contribution to the structure plus the ordering generated by the long tail in the molecule stabilizes the system. The properties and structure of the lamellar structure are given in Table 3.5.

The lamellar phase can be optically positive or negative uniaxial; the sign may change with temperature change. The common texture observed with these phases is the focal conic and its variation, the fan texture. The optical axis is parallel to the long axis of the molecules in the layers. A pseudoisotropic texture that is dark between crossed polarizers is often interspersed with bright birefringent bands (oil streaks).

The structure of the lamellar phase can be determined by X-ray diffraction. The thickness of the layers is generally less than twice the length of the molecule. The thickness decreases with increase of temperature and with increase of water concentration; these changes of thickness may be attributed to the fold of the hydrocarbon chains and/or the tilt of the molecules in the layers. Addition of more water to the lamellar structure may result in a cubic structure.

2. Cubic Structure

The cubic structure is optically isotropic. It is generated by the amphiphilic molecules packing in spheres, and then the spheres in an aqueous environment packing in a cubic pattern. One kind of spherical packing has the ionic portion of the molecule on the surface of the sphere and the or-

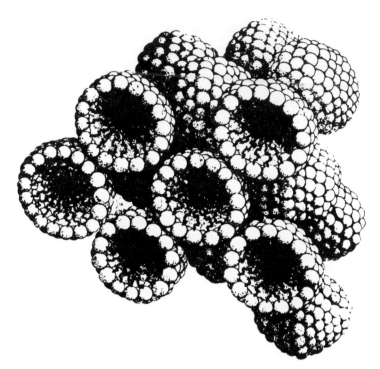

Fig. 3.6 Schematic representation of cubic packing of spheres. The spheres are generated by interaction of amphiphilic molecules and water. Modified with permission from Rosevear (1968).

ganic portion (water insoluble) of the molecule directed to the center of the sphere. This is a low-energy structure with the organic portion of the molecules dissolving in each other. The ionic units on the surface of the sphere interact with water and generate a cubic packing. As pointed out, this system is classified as liquid crystalline. A schematic representation of this structure is given in Fig. 3.6. The characteristics of the cubic structure are outlined in Table 3.6. In an inverted cubic structure the ionic portion of the molecule is addressed to the center of the sphere and the organic "tail" is projected into the surrounding environment. The sphere would have water in its interior.

3. Hexagonal Structure

Adding more water to the cubic structure may result in the formation of a hexagonal structure. In the regular structure the organic portion of the molecule is addressed to the center of the cylinder and the ionic portion of the molecule is located on the circumference of the cylinder. The cylinders

TABLE 3.6

Cubic Packing of Lyotropic Liquid Crystals[a]

Designation	Optical properties	Basic structure	Description of proposed structure
Optical isotropic meso-phase, normal viscous isotropic phase	Isotropic	Body centered or face centered	Packing of spheres
Optical isotropic meso-phase, reversed viscous	Isotropic	Body centered	Packing of spheres

[a] (1) Phases characterized by three-dimensional periodicity. (2) Structural arrangement displaying cubic symmetry.

then interact with water to generate the hexagonal structure. The cylinders pack with their long axes lying parallel to each other. The hexagonal structure is sketched in Fig. 3.7 and a description of the properties of the structure are given in Table 3.7. An inverted version of this structure (see Figs. 3.4f, g) can be represented with the ionic heads directed to the center of the cylinder with water filling the center of the cylinder. The organic portion of the molecule is addressed to the outer surface of the cylinder.

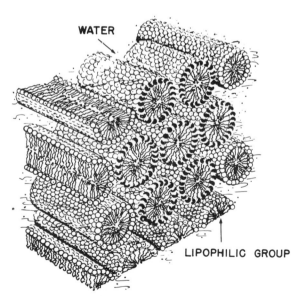

WATER

LIPOPHILIC GROUP

Fig. 3.7 Schematic representation of the packing pattern of rodlike particles. This sketch illustrates hexagonal packing of amphiphilic molecules with water. Modified with permission from Rosevear (1968).

4. Micellar Structures

Micelles are large polynuclear aggregates in solutions. The aggregates have different shapes. They are thermodynamically stable because of intermolecular interactions. Micellar solutions are not considered to be liquid crystals. The two most common geometries of micelles are spherical and cylindrical in shape. The ions on the surface of the sphere are hydrophilic groups directed toward water (Fig. 3.6). Inverted micelles in a hydrocarbon environment have their polar groups directed to the center of the sphere. These inverted micelles can trap water which dissolves the polar groups. The molecular packing in cylindrical design can be regular with the hydrocarbon part of the molecule addressed to the center of the cylinder with the ion portions of the molecules lying on the surface of the cylinder. The inverted cylindrical micelles have their polar groups directed to the center of the cylinder. The difference in size of the molecular cluster evidently determines whether the structure is a liquid crystal or a micelle. The micellar structure is isotropic while liquid crystals are anisotropic (except smectic D).

The transition from micellar to homogeneous solution is gradual and not as sharp as or as well defined as the other lyotropic phase changes.

5. Gel Phase

On lowering the temperature of some lipid–water systems, the ordinary mesophases are transformed to a transparent phase—a "gel." The gel structure is an intermediate state between the liquid crystalline state with its aggregates of unordered hydrocarbon chains in a semiliquid state and the crystalline state with completely ordered chains. On further cooling

TABLE 3.7

Hexagonal Packing of Lyotropic Liquid Crystals[a]

Designation	Optical properties	Basic structure	Description of proposed structure
Middle phase	Anisotropic	Two-dimensional hexagonal	Long, mutually parallel rods in hexagonal array
Middle phase	Anisotropic	Two-dimensional hexagonal	Amphiphilic molecules in rods are essentially in radial pattern
Hexagonal complex phase, normal	Anisotropic	Two-dimensional hexagonal	Indefinitely long, mutually parallel rods in hexagonal array

[a] (1) Structural arrangement displaying Bragg spacing ratio $1:1/3:1/4:1/7$. (2) Two-dimensional periodicity. (3) Molecules packed in rodlike pattern.

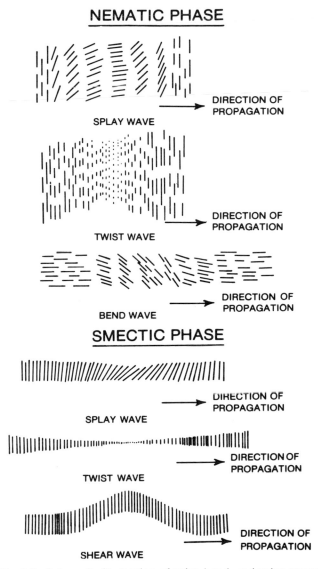

Fig. 3.8 Schematic illustration of twist, bend, and splay waves.

the gel phase is converted to the crystalline state. The gel phases are thermodynamically stable phases with very special structural properties. The characteristics which the gel phases possess make them as interesting as the lamellar liquid crystals from a biological point of view. Cholesteryl sulfate and cholesteryl monophosphate both form liquid crystals and gels with water.

IV. REMARKS

A few comments on the potential role of liquid crystals in biochemical reactions may be of interest. The liquid crystalline state can be a substrate on which a chemical reaction can take place. For example, in the self-ordering nematic structure, the long axes of the molecules are essentially parallel and as they line up they generate a "layer" of molecules which can serve as a substrate on which reactant molecules can interact. Such surfaces may be used in simple organic reactions such as isomerization, and in biological processes such as enzymatic oxidation, reduction, and dehydrogenation.

As pointed out earlier in this chapter, the liquid crystalline state possesses one-dimensional order (nematic) and two-dimensional order (smectic). These structures respond readily to energy changes. Figure 3.8 illustrates schematically torque waves in nematic and smectic liquid crystals. The splay-wave pattern can explain the transfer of ions and molecules across the liquid crystalline structure through the spread of molecules in the liquid crystal. The intrusion of chemicals into a liquid crystal could change the pitch of the helix thus changing the optical properties of the liquid crystal. The bend structure might readily explain the curvature properties of cell membranes.

REFERENCES

Brown, G. H., ed. (1976). "Advances in Liquid Crystals," Vol. 2. Academic Press, New York.

Chandrasekhar, S. (1977). "Liquid Crystals." Cambridge Univ. Press, London and New York.

de Gennes, P. G. (1974). "The Physics of Liquid Crystals." Oxford Univ. Press (Clarendon), London and New York.

de Vries, A. (1973). A new classification system for thermotropic smectic phases. *Mol. Cryst. Liq. Cryst.* **24**, 337.

de Vries, A. (1975). X-Ray studies of liquid crystals: V. Classification of thermotropic liquid crystals and discussion of intermolecular distances. *Pramana Suppl.* **1**, 93.

de Vries, A. (1979). Two classification systems for smectic phases based on symmetry and order. *J. Chem. Phys.* (in press).

Friberg, S., ed. (1976). "Lyotropic Liquid Crystals," Adv. Chem. Ser. No. 152. Am. Chem. Soc., Washington, D.C.

Rosevear, F. B. (1968). Liquid crystals: The mesomorphic phases of surfactant compositions. *J. Soc. Cosmet. Chem.* **19**, 581.

Sackmann, H., and Demus, D. (1966). The polymorphism of liquid crystals. *Mol. Cryst. Liq. Cryst.* **2**, 81.

Sackmann, H., and Demus, D. (1973). The problems of polymorphism in liquid crystals. *Mol. Cryst. Liq. Cryst.* **21**, 239.

Chapter 4
Optical Properties of Liquid Crystals

I. OPTICAL CHARACTERISTICS

A. Introduction

The optical properties of the four states of matter may be classified as isotropic or anisotropic. The term isotropic means there are equal properties along the x, y, z axes in Cartesian coordinates.

Liquid crystals are optically anisotropic (i.e., transmit light waves of different velocities in different directions). Such liquid crystals are double refractive or birefringent. For an excellent review of optical properties, see Hartshorne (1974).

Natural light travels in a straight line from an object to the eye. It vibrates in all directions along the line of propagation. In studying the microscopic properties of liquid crystals we are most often concerned with polarized light. The term polarization has been used whenever anything has a property in one direction that it does not have in the other. The polarization of natural light can be accomplished in several ways, but we will mention only the use of the Nicol prism. Two prisms are needed for the polarizing microscope.

When two Nicol prisms are placed in series, polarized light is transmitted through them. However, if either is rotated relative to the other through 90° (crossed polarizers) the light will fail to pass. When a liquid crystal is observed through crossed polarizers, intense bands of color are seen.

A homogeneously aligned specimen of a nematic liquid crystal is optically uniaxial positive and strongly birefringent. The molecules lie parallel to the plane and are said to be homogeneous or planar. If the long axis of the molecules lies perpendicular to the plane of the preparation, the structure is homeotropic.

Anisotropy may be uniaxial or biaxial in liquid crystalline structures.

INCIDENT LIGHT

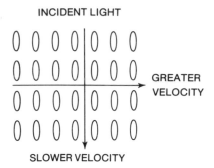

Fig. 4.1 Light transmission through a smectic structure. Velocity of light transmitted perpendicular to layers is less than that transmitted parallel to layers. The ellipses represent molecules.

For uniaxial structures the light vibrating in one direction travels at a different rate than light vibrating in a direction perpendicular to it. In many smectic and nematic liquid crystals the velocity of light transmitted perpendicular to the layers of the molecules is less than that transmitted parallel to the layers (Fig. 4.1). Materials showing this property are said to be optically positive. The cholesteric–nematic structure behaves like a negative uniaxial crystal; that is, the velocity of light vibrating perpendicular to the molecular layers has maximum velocity. When the velocities of light are equal in two different directions in a liquid crystal, the crystal is biaxial.

B. Birefringence

When ordinary light, which vibrates in all directions, strikes the surface of a birefringent material, the beam is broken into two polarized components which vibrate at right angles to each other and travel at different speeds. These components have different angles of refraction and are parallel (Fig. 4.2). One way to identify the characteristics of a liquid crystal is to test it for birefringence. Birefringence is characteristic of crystals and liquid crystals. A beam of white light entering the liquid crystal divides into two beams which, refracted at different angles, are emitted parallel to each other (Brown, 1967). The two emitted beams are polarized and vibrate at right angles to each other (Fig. 4.2).

C. Dichromism

Many liquid crystalline materials show dichromism. They allow one component of polarized light to be absorbed more than the other. Of the liquid crystalline materials, the cholesteric–nematic structure exhibits the most interesting dichroic properties. When white light strikes the surface of a cholesteric substance, it is separated into two components, one rotat-

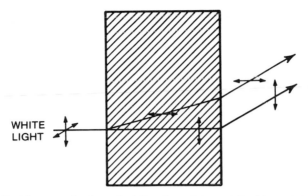

Fig. 4.2 Birefringence is characteristic of crystals and liquid crystals. A beam of white light entering the material divides into two beams which, refracted at different angles, are emitted parallel to each other. The two emitted beams are polarized and vibrate at right angles to each other.

ing clockwise and the other counterclockwise (Fig. 4.3). Depending on the type of liquid crystal, one component reflected from the surface is of one color, while the other component is transmitted and is of another color. When illuminated with white light, cholesteric substances show a characteristic iridescence.

D. Optical Activity

Cholesteric–nematic liquid crystals have unusual optical activity, that is, the ability to change the direction of vibration of polarized light. This property is greater for cholesteric–nematic liquid crystals than that of any other known substances. For example, a 1-mm section of quartz will rotate the polarization plane of blue light some 39°, whereas a 1-mm section of isoamyl-p-(4-cyanobenzylideneamino)cinnamate, a cholesteric–nematic liquid crystal, will rotate the polarization plane hundreds of complete turns (Hartshorne 1974). The direction and angle of rotation for a unit path length are characteristic for each substance.

II. TEXTURES OF THERMOTROPIC LIQUID CRYSTALS

A. Introduction

The microscopic observation of liquid crystals in linearly polarized light is widely used (Sackmann and Demus, 1963, 1973; Saupe, 1977). The textures observed in microscopic studies are valuable in the scheme of classification of liquid crystals. Combined with miscibility studies, differential thermal analysis and X ray, a system of classification of liquid crystals has evolved.

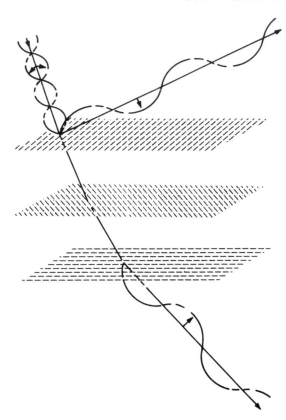

Fig. 4.3 Circular dichroism of cholesteric liquid crystals. Unpolarized light hitting the surface of the material divides into two components, one reflected, the other transmitted. One component has one electric vector rotating clockwise and the other counterclockwise. This property imparts iridescence to the structure when illuminated with white light.

Textures of liquid crystals are usually observed with the liquid crystal placed between a glass slide and a coverslip in polarized light (crossed polarizers). One may at times study texture characteristics by removing the coverslip. The pictures shown in this chapter are selected to show observable texture characteristics. It should be mentioned that in practice, variations in textures may be observed. These variations depend on the properties of the substances and the manner in which the sample was prepared. Often textures appear whose characteristics are somewhat obscure because paramorphs are formed due to impurities and/or wall effects. The name paramorph is given to a crystal or liquid crystal whose internal structure has changed to that of a different form without producing any change in external form. Paramorphism is often observed after transition of one liquid crystalline modification to another. The texture of the original form is only slightly altered. Paramorphism often occurs with smectic structures. For example, in the transition of a smectic B to a smectic A, the smectic B may appear in the form of a fan-shaped texture which is characteristic of smectic A.

A few selected textures of thermotropic liquid crystals will be presented here.

B. Schlieren (Streaks) Textures

Ordinary nematic liquid crystals often show the schlieren textures (Fig. 4.4), especially in layers, 0.1 mm or greater in thickness. The texture results from a nonhomogeneous orientation of the molecules.

The films show a large number of threadlike disclinations (threads) in the structure. In thinner films, a schlieren texture with pointlike singularities can be obtained. The optical characteristics of this texture are dark brushes which start with point singularities. The singularities are vertical threads and are characterized by the number of dark brushes which emerge from them between crossed polarizers. The point singularities are characterized by $|S|$ = number of brushes/4. Commonly, one finds points with 4 or 2 dark brushes. The sign of the point is positive when the brushes turn in the same direction the polarizers are rotated and negative if they turn in the opposite direction. Point singularities with $S = +\frac{1}{2}, -\frac{1}{2}, +1, -1$ have been observed in nematic liquid crystals. The point singu-

Fig. 4.4 Nematic schlieren texture, crossed polarizers, 120×. Note 2 and 4 dark brushes.

larities are projections of threads that are oriented with their long axes perpendicular to the surface. In thicker samples (>0.1 mm) the ends of the threads are each attached to the walls and otherwise float freely. These threads are visible as straight or bent lines.

The schlieren texture is also found in other liquid crystals such as smectic A and smectic B modifications.

C. Cholesteric–Nematic Textures

The cholesteric–nematic liquid crystal exhibits optical properties that are striking and different from other liquid crystals. One way to observe these optical properties is to prepare a sample between a glass microscope slide and a cover glass with the optical axis of the cholesteric molecular unit perpendicular to the surfaces that bound it. This type of orientation is generally generated spontaneously on preparation of the sample or can be generated by adjusting the cover glass slightly. Such a system shows a negative optical character. There is strong optical activity along the optic axis.

Cholesteric–nematic liquid crystals have the same orientational order as ordinary nematic liquid crystals but differ in texture and molecular arrangement. The alignment of the molecules in the planar texture is normal to a vertical axis and uniformly parallel in horizontal planes. The alignment direction turns linearly with respect to the vertical axis. This alignment gives a helical structure. The degree of twist is characterized by the pitch of the helix, the distance for a 2π turn.

An irregular variation of the planar texture is known as a focal conic texture. Using thin films, the focal conic texture can be converted to a planar texture by a simple mechanical twist, such as a shift of the cover slide. The texture transfer can take place in some cases by using ac or dc electrical fields. This is especially the case for cholesteric structures generated by adding an optically active compound to a nematic liquid crystal.

When a cholesteric–nematic liquid crystal is placed on a microscope slide and the analyzer is rotated, a color change will occur. The rotation of the light changes sign at a wavelength λ_0 which varies with the kind of stance and the temperature. Within a narrow band around λ_0 one component of the circularly polarized light will be reflected and the other component of the opposite band will be transmitted. A cholesteric–nematic liquid crystal observed under diffused daylight scatters the light in different directions producing a striking display of colors in the visible region.

Spontaneously twisted cholesteric–nematic liquid crystals all have asymmetric molecules. This type of twisted nematic liquid crystal is distinguished from one that is converted to the twisted type by a mechanical process. Cholesteric–nematic liquid crystals exhibit both *dextro* (*d*) and *levo* (*l*) types. Mixing solutions of the *d* and *l* types, one can prepare a racemic mixture that is optically inactive. This result was first observed by Friedel (1922) when he found that by mixing *d* and *l* forms he could destroy the optical activity and convert the system to an ordinary nematic structure.

We will mention one other optical property of cholesteric–nematic structures. Grandjean found that a cholesteric-type liquid crystal, when melted into the cleavage crack in a sheet of mica, forms a series of regularly spaced bands separated by sharp lines. The bands follow the contour of the crack and the separation between the bands corresponds to a constant change in the thickness of the crack.

D. Focal Conic Textures

The focal conic texture exhibits two modifications, the fan-shaped texture (Fig. 4.5) and the polygonal texture (Fig. 4.6). The fan-shaped texture shows disclination lines forming hyperbolae, and the polygonal texture forms elipses. The smectic A structure is an example of a simple focal conic texture. There are some similarities between the smectic textures

Fig. 4.5 Simple fan-shaped texture of smectic A, crossed polarizers, 120×.

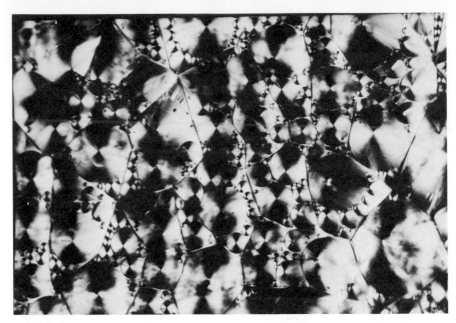

Fig. 4.6 Simple polygon texture of smectic A, crossed polarizers, 120×. (Courtesy of H. Sackmann.)

Fig. 4.7 Mosaic texture of smectic B, crossed polarizers, 320×. (Courtesy of Mary Neubert.)

and cholesteric–nematic textures, since both phases in their equilibrium structures have a translational periodicity. This periodicity is one-half the pitch in cholesteric–nematic structures and the layer distance in the smectic structures. We know that the layer distance in smectic liquid crystals remains essentially constant, while the cholesteric pitch varies readily, which gives the cholesteric textures less regular geometric features than the smectic textures. The focal conic texture is exhibited by other liquid crystals including smectic C and smectic F.

E. Mosaic Textures

Smectic liquid crystals B, E, and G all exhibit mosaic textures. A mosaic texture has different-colored, optically homogeneous regions with irregular boundaries. In each region the molecular arrangement is uniform and the optical axes in different regions have different orientations (Fig. 4.7). The optical homogeneity of the regions indicates the layers of these textures are planar.

F. Isotropic Texture

Liquid crystals with cubic packing of spheres generated by clusters of molecules are optically isotropic. Among liquid crystal structures the smectic D modification is an isotropic texture and is stable.

III. TEXTURES OF LYOTROPIC LIQUID CRYSTALS

A. Introduction

Microscopy of lyotropic liquid crystals has not been studied as extensively as the thermotropic structures. Polymorphism of lyotropic liquid crystals was discussed in Chapter 3. In this section we shall consider the microscopy of amphiphilic compounds. Lyotropic liquid crystals can also exhibit thermotropic properties. For example, if a lyotropic liquid crystal, which shows birefringence, say, at room temperature, is heated, the birefringence will disappear and the system will become isotropic; but the birefringence will reappear on cooling.

B. Textures of Lyotropic Liquid Crystals

The publications of Rosevear (1954, 1968) are very useful for those who want to know more about these textures and others of the lyotropic kind. The neat phase (G) is much more fluid than the middle phase. If a tube

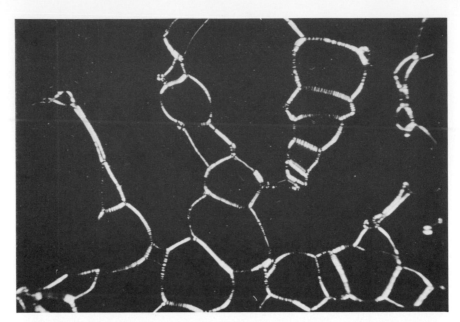

Fig. 4.8 Pseudo-isotropic texture with oily streaks, lamellar phase; crossed polarizers, lecithin (vegetable) and water. From Saupe (1977), *J. Colloid Interface Sci.* **58**, 549. Reprinted by permission.

Fig. 4.9 Fan texture in middle soap; 200×, crossed polizers, 64°C, decylammonium chloride and water. From Saupe (1977), *J. Colloid Interface Sci.* **58**, 549. Reprinted by permission.

containing a sample of the neat phase is inverted, it will flow readily under its own weight. A number of modifications of textures of neat phases are known. We show only one of these in Fig. 4.8.

The middle phase (M_1), even though it has a higher water content than the neat phase, is much stiffer. The middle phase does not flow under the influence of gravity; it does flow plastically if subjected to sufficient external force. In Fig. 4.9 we show the fanlike texture of the middle phase based on the packing of the cylindrical molecular aggregates. The M_2 phase has a structure like the M_1 phase but with the polar groups directed inward with the hydrocarbon tail projecting outward.

A convenient way to study a lyotropic liquid crystal system is to place a drop of the dilute isotropic solution between a microscope slide and a cover glass and observe the sequence of mesophases as the water slowly evaporates.

The optical sign of the G phase is almost always positive, while most M phases are negative.

Now that we have discussed the classification of liquid crystals, their optical properties, and their structure, the next chapter will be a discussion of whether the molecules that we associate with life itself possess similar structures and behavior properties.

REFERENCES

Brown, G. H. (1967). Liquid crystals. *Chemistry* **40**, 10.

Friedel, G. (1922). The mesomorphic state of matter. *Ann. Phys.* (*Leipzig*) [4] **18**, 273.

Hartshorne, N. H. (1974). *In* "Liquid Crystals and Plastic Crystals," Vol. 2 (G. W. Gray and P. A. Winsor, eds.), pp. 24–61. Wiley, New York.

Rosevear, F. B. (1954). The microscopy of the liquid crystalline neat and middle phases of soaps and synthetic detergents. *J. Am. Oil. Chem. Soc.* **31**, 628.

Rosevear, F. B. (1968). Liquid crystals: The mesomorphic phases of surfactant compositions. *J. Soc. Cosmet. Chem.,* **19**, 581.

Sackmann, H., and Demus, D. (1963). Isomorphiebeziehungen zwischen kristallin-flüssigen Phasen. *Z. Phys. Chem.* **222**, 127.

Sackmann, H., and Demus, D. (1973). The problems of polymorphism in liquid crystals. *Mol. Cryst. Liq. Cryst.* **21**, 239.

Saupe, A. (1977). Textures, deformation and structural order of liquid crystals. *J. Colloid Interface Sci.* **58**, 549.

Chapter 5
The Structural Molecules of Life

I. INTRODUCTION

The main substances of living cells are water, certain inorganic salts, and organic compounds. These organic molecules consist of relatively few atoms: carbon, hydrogen, oxygen, and nitrogen. Complex organic compounds, the proteins, the lipids, the carbohydrates, and the nucleic acids, for the most part, are the essential molecules of life. It is of interest to review briefly the structure and properties of these molecules that are considered essential to life before exploring how these molecules are structured in the cell. Later, we can ask whether or not such structures form and behave like liquid crystalline systems. Let us then examine the structure and properties of a few of the molecules to which we will be referring in our discussion of living systems.

II. PROTEINS

The most common feature of all living organisms is that proteins are essential for life. They function to regulate metabolic processes, as catalysts for biological reactions, and play a very important part in their structure. The proteins are the largest and most complex molecules known. The units of which their structures are built consist of about 20 different amino acids (Fig. 5.1). Amino acids are joined together by peptide bonds. The peptide linkage is planar and rigid (**III**). A large number of amino acids joined together through such peptide linkages are called *polypeptides*. The amino acids are strung together in chains hundreds to thousands of units long, in different proportions, in all types of sequences, and with a great variety of branching and folding. An infinite number of different proteins is possible, for no two species of organisms possess exactly the same proteins.

Proteins vary in molecular weight from around 5000 to the order of mil-

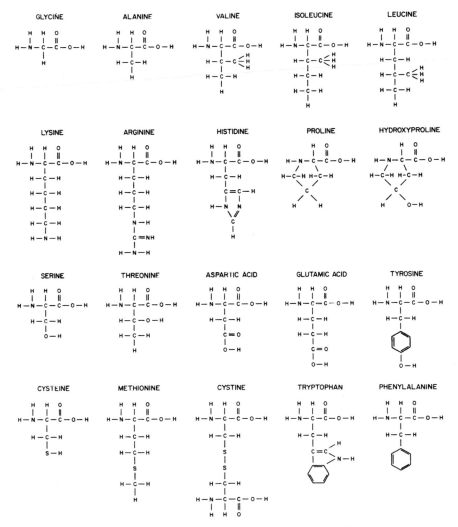

Fig. 5.1 Structure of representative amino acids.

lions. Proteins are generally divided into the fibrous, which are composed of elongated filamentous chains, and the globular, with a considerable amount of folding of the long polypeptide chain. X-Ray studies of native proteins and synthetic polypeptides led Linus Pauling to propose a structure of greatest stability—which is the α-helix (Pauling, 1960). The α-helix has a spiral chain of repeating amino acids held together by hydrogen

$$
\begin{array}{c}
O \quad\quad\quad H \\
\| \quad\quad\quad | \ \ R \\
C \quad\quad\quad C_\alpha \\
\diagup C_\alpha \ \ \diagdown \quad N \diagup \\
R \quad H \quad\quad | \\
H
\end{array}
$$

(III)

bonds (Fig. 5.2). The helix contains about four amino acid residues for each full turn of the spiral. The α-helix is a one-dimensional subcrystalline arrangement. Another type of structure is the β-configuration (Pauling, 1960). Here, two or more peptide chains are tied together laterally by hydrogen bonding. Wherever hydrogen bonding occurs, a crystalline structure is observed. This hydrogen bonding exists when the crystal is dissolved in water to form the liquid crystal.

A series of classical studies of polypeptides as a liquid crystalline system was carried out by Robinson (1956, 1958, 1966), in which the properties of the polypeptide, poly-γ-benzyl-L-glutamate (PBLG) (IV) in different solvents were studied.

In concentrated solutions, the polypeptide PBLG is birefringent. The solutions which show birefringence develop parallel and equally spaced narrow bands which are alternately bright and dark. The spacing between adjacent dark or bright bands is periodic and varies from 2 to 100 μm, which is dependent on the concentration, the solvent, and the temperature. The observed periodicity is relatively independent of the molecular weight of the polymer.

Microscopic observations show that the solute has a twisted structure with a pitch that can be observed optically in the visible region of the spectrum. If the concentration of the glutamate PBLG is increased, the pitch of the helix decreases. This decrease in pitch can continue until the system looks like a cholesteric–nematic liquid crystal.

Solutions containing equal concentrations of the l-PBLG and the dextro enantiomorph (mirror image forms) show no optical activity and have all

$$
\left[
\begin{array}{l}
CO \\
| \\
CH{-}CH_2{-}CH_2{-}COOCH_2C_6H_5 \\
| \\
NH
\end{array}
\right]_x
$$

(IV)

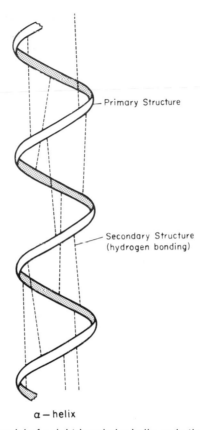

Primary Structure

Secondary Structure
(hydrogen bonding)

α − helix

Fig. 5.2 Schematic model of a right-handed α-helix, as in the structure of proteins.

the properties of a nematic liquid crystal. For example, mixing solutions of equal concentrations of PBLG in methylene chloride and in 1,4-dioxane also gives a nematic liquid crystalline structure. The methylene chloride solutions are dextro rotatory and the 1,4-dioxane solutions are levo rotatory.

Evidence shows that the polypeptide chain of PBLG is wound into the α-helix structure (Fig. 5.2). The structure has adjacent coils bound together by hydrogen bonds formed between neighboring —C=O— and NH groups that are four peptide bonds apart. The side chains, —CH_2CH_2—$COOCH_2C_6H_5$, project radially from the helix, as illustrated in Fig. 5.3. The polymer that is formed behaves like a stiff rod with projecting side units. These rods may pack in different ways to form liquid crystals with the most likely geometry being hexagonal. Other polypep-

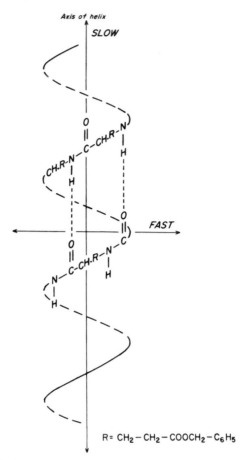

Fig. 5.3 The α-helix structure applied to poly-γ-benzyl-L-glutamate (PLBG) showing hydrogen bonds between adjacent coils.

tides also show these properties. However, the role of liquid crystals in protein chemistry has hardly been explored.

III. LIPIDS

The lipids are another important structural molecule, and the widespread presence of lipids in plants, animals, and microorganisms has led to considerable experimental studies of their structural and metabolic roles in the cell.

Lipids are a rather heterogeneous class of compounds. They are classified as neutral lipids, phosphatides and sphingolipids, and glycolipids and

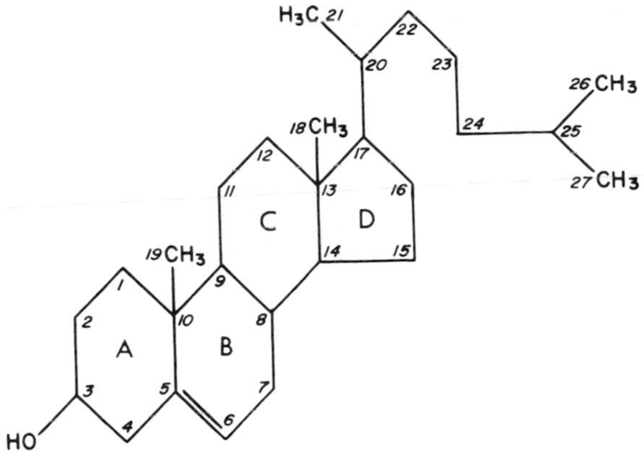

Fig. 5.4 Cholesterol structure.

terpenoid lipids, including carotenoids and steroids. Simple lipids are esters of fatty acids and an alcohol. Cholesterol (Fig. 5.4) and fatty acids form esters which are found in living systems. These esters show liquid crystalline properties.

Phospholipids, and in particular lecithin, phosphatidyl choline, cephalin, and phosphatidyl ethanolamine, are the most abundant of the naturally occurring phosphoglycerides. Their basic structure is illustrated in Fig. 5.5. The most common phospholipids have chain lengths (represented by R_1 and R_2 in Fig. 5.5) 16–20 carbons, with up to four carbon–carbon double bonds.

Phospholipids, fatty acids, and cerebrosides are examples of what may be classed as polar lipids. They possess strongly polar or charged groups arranged in such a way in the molecule that they may be oriented toward water or other polar molecules, and their nonpolar portions are oriented away from the polar environment (Fig. 5.6a). These lipids have melting points of 200°–300°C. Lipids that have no polar groups in the molecule melt at much lower temperatures, around 70°C.

Most phospholipids disperse molecularly in water to only a small extent, and if larger quantities of phospholipids are introduced into the aqueous medium, aggregates called micelles and/or liquid crystals are formed. Whether one forms liquid crystals or micelles depends primarily on the composition and temperature of the system.

For example, lecithin, when mixed with water, undergoes phase transition (depending on the temperature) from a turbid to a clear solution, and it exhibits the properties of a liquid crystal. When lecithin in water is ob-

Phospholipid Structure

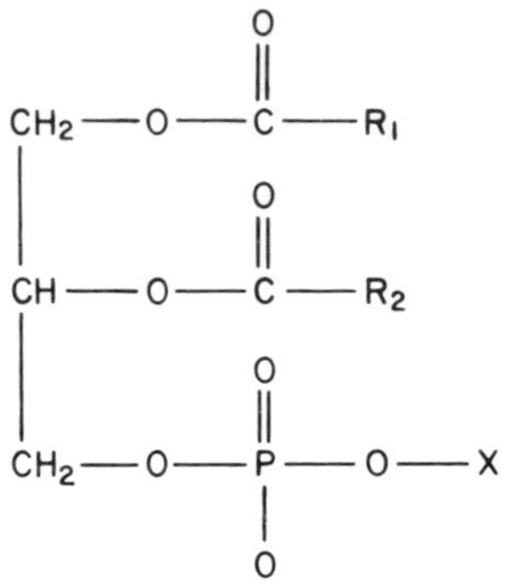

$$
\begin{array}{l}
\quad\quad\quad\quad\quad\quad\quad O \\
\quad\quad\quad\quad\quad\quad\quad \| \\
CH_2\!-\!O\!-\!C\!-\!R_1 \\
| \\
\quad\quad\quad\quad\quad\quad\quad O \\
\quad\quad\quad\quad\quad\quad\quad \| \\
CH\!-\!O\!-\!C\!-\!R_2 \\
| \\
\quad\quad\quad\quad\quad\quad\quad O \\
\quad\quad\quad\quad\quad\quad\quad \| \\
CH_2\!-\!O\!-\!P\!-\!O\!-\!X \\
\quad\quad\quad\quad\quad\quad\quad | \\
\quad\quad\quad\quad\quad\quad\quad O
\end{array}
$$

Phospholipids	X
Phosphatidyl choline (Lecithin)	$-CH_2-CH_2-\overset{+}{N}\big\langle\begin{smallmatrix}CH_3\\CH_3\\CH_3\end{smallmatrix}$
Phosphatidyl ethanolamine (Cephalin)	$-CH_2-CH_2-\overset{+}{N}H_3$
Phosphatidyl serine	$-CH_2-\underset{\underset{NH_3^+}{\|}}{CH}-COO^-$
Phosphatidyl inositol	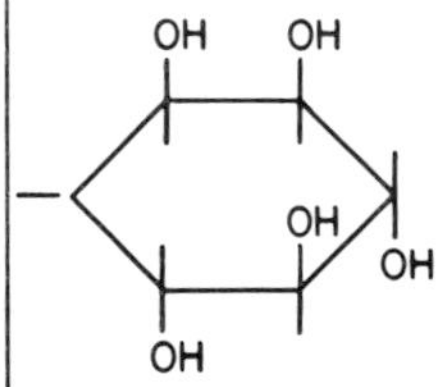 $-$ inositol ring with OH, OH, OH, OH, OH

Fig. 5.5 Basic structure of common phospholipids. R_1 and R_2 are akyl radicals with chain lengths of 16 to 20 carbons.

served under a microscope, it is seen to form lamellae (Fig. 5.7) or myelin figures similar to those that are observed in living cells (Figs. 7.5a and 7.9; see also Figs. 10.5 and 10.6).

In the formation of liquid crystals, there are many areas that we do not understand. But there are many biological implications of liquid crystal formation, particularly those of the phospholipids and the biological sur-factants, which are of considerable physiological importance. Most im-

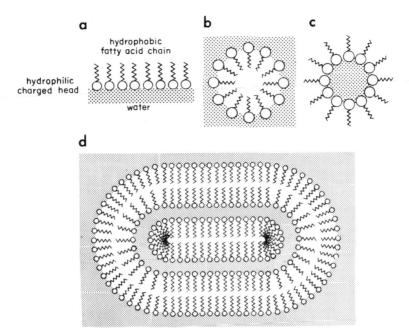

Fig. 5.6 (a) A monolayer of phospholipid molecules in water. The phospholipids are symbolized by a circle representing the charged hydrophilic end, and the zig-zag line represents the hydrophobic fatty acid chain. (b) If the liquid is polar, like water, the charged phosphates face outward. (c) If it is nonpolar, like benzene, they face inward. It can also exist as a combination of (b) and (c) as indicated in (d).

portant is their relation to the structure of the cell membrane and to many other cellular structures (Chapman, 1973). These structural relationships will be pointed out in Chapters 6 through 11.

IV. CAROTENOIDS

The terpenoid lipids include the carotenoids, which are the most abundant pigments found in nature. The carotenoids are easily and abundantly synthesized by plants, and their biosynthesis is generally associated with the 20-carbon, aliphatic alcohol phytol (Fig. 5.8). Carotenoids are divided into two main groups, the hydrocarbon carotenes, as represented by carotene, $C_{40}H_{56}$, and the oxygen-containing derivatives, the xanthophylls, such as lutein (Fig. 5.8). The carotenoid molecule is made up of a chromophoric system of alternating single and double interatomic linkages—called a polyene chain of conjugated double bonds—between the carbon

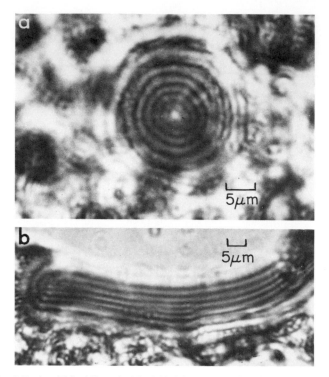

Fig. 5.7 Lecithin (dipamityl lecithin) in water: (a) 1500×, (b) 1000×.

atoms. The spectral characteristics, and therefore the color of the carotenoids, are largely determined by the number of conjugated double bonds in the molecule. The large number of these conjugated double bonds offers the possibility of either trans or cis geometric configurations (Figs. 5.8 and 5.9).

An important change occurred during the evolutionary development of animals—they became dependent on the ingestion of plants for their source of carotenoids. It is not the ingested plant carotenoids themselves but their derivative, vitamin A, that is necessary for all animal life (Fig. 5.9). Thus, the carotenoids can be seen to play a central role in the biochemical evolution from the plant C_{40} (β-carotene) to animal C_{20} (vitamin A).

We will elaborate further on the importance of the carotenoids to visual excitation and in the development of photoreceptor structures in Chapter 9.

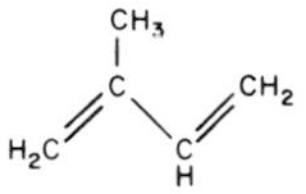

ISOPRENE

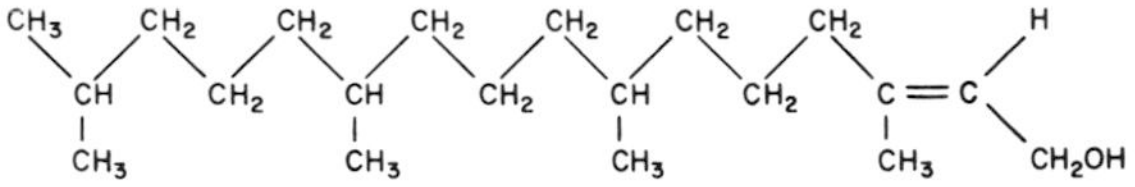

PHYTOL

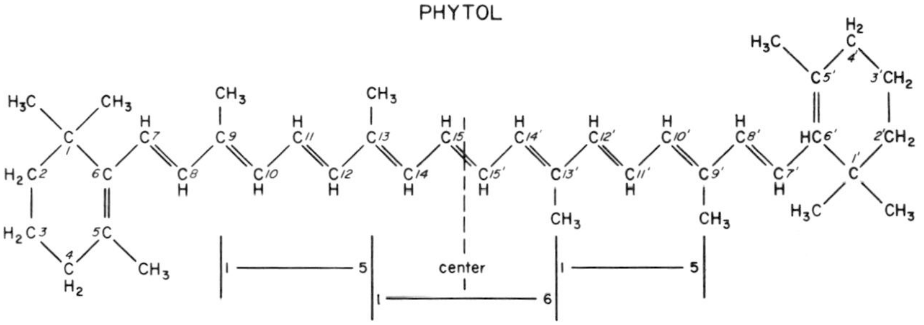

β-CAROTENE

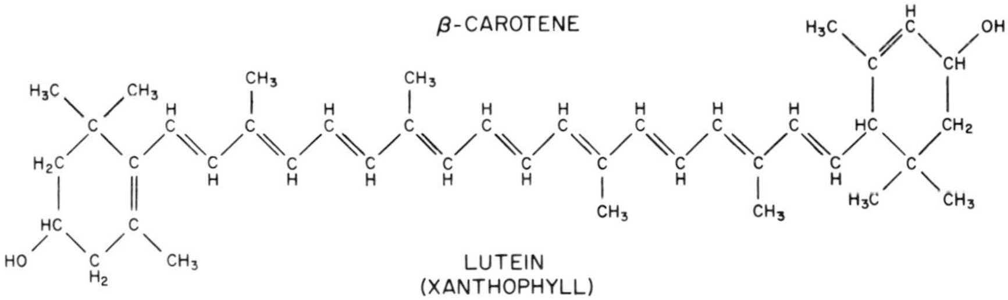

LUTEIN
(XANTHOPHYLL)

Fig. 5.8 The carotenoid structure.

V. POLYSACCHARIDES

The carbohydrates provide an important source of energy for living organisms, a means for storing energy, and some carbohydrates also function as structural molecules for the cell as well. Carbohydrates are comprised of monosaccharides, disaccharides, and polysaccharides. The monosaccharides are classified in accordance with the number of carbon atoms in the molecule. The pentoses ribose and deoxyribose are found in the molecules of nucleic acids (Fig. 5.10).

Polysaccharides $(C_6H_{10}O_5)_n$ are condensations of many molecules of monosaccharides. The most important polysaccharides to living organisms are starch, glycogen, cellulose, and chitin (Fig. 5.11). In many

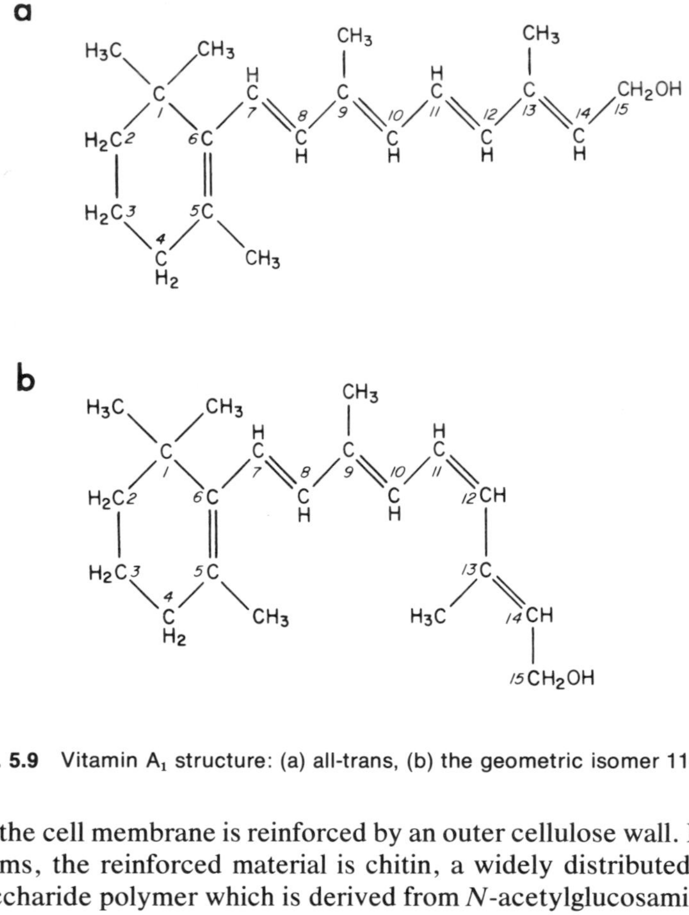

Fig. 5.9 Vitamin A₁ structure: (a) all-trans, (b) the geometric isomer 11-cis.

plants, the cell membrane is reinforced by an outer cellulose wall. In other organisms, the reinforced material is chitin, a widely distributed animal polysaccharide polymer which is derived from N-acetylglucosamine. The function of chitin is essentially a structural one and forms the bulk of the exoskeleton or cuticle, the hard outer covering of insects and crustaceans. Polysaccharides with water can also form liquid crystals.

VI. NUCLEIC ACIDS

Nucleic acids are essential in all living organisms to affect the processes of reproduction, growth, and differentiation. The nucleic acids are DNA (deoxyribonucleic acid) and RNA (ribonucleic acid). Both DNA and RNA are long chains of alternating sugar and phosphate groups. In DNA the sugar is deoxyribose; that is, the carbon at the 2′-position carries simply

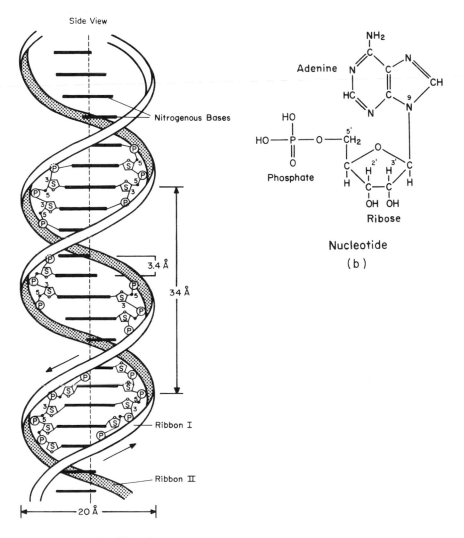

Side View

Nitrogenous Bases

3.4 Å

34 Å

Ribbon I

Ribbon II

20 Å

DNA Molecular Structure

(a)

Adenine

Phosphate

Ribose

Nucleotide

(b)

Fig. 5.10   The structure of the DNA molecule, that of a double helix (a). This is a representation of the Watson–Crick model of the molecule (hydrated "B" form). From Etkin (1973), *Bioscience* **23**, 653. Reproduced with permission. (b) A nucleotide structure, composed of a nitrogenous base adenine, a sugar riboαe (or deoxyribose), and a phosphate.

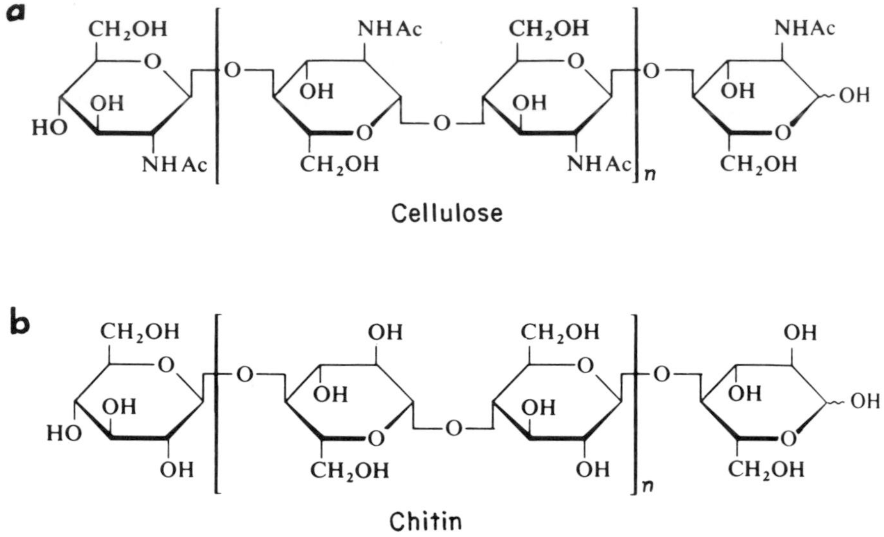

a

CH₂OH · · · NHAc · · · CH₂OH · · · NHAc

Cellulose

b

CH₂OH · · · OH · · · CH₂OH · · · OH

Chitin

Fig. 5.11 Polysaccharide polymers: (a) cellulose, (b) chitin.

hydrogen, and in RNA the sugar is ribose, because the 2′ carbon carries a hydroxyl group. Each purine or pyrimidine, with its sugar and phosphate, is referred to as a nucleotide (see Fig. 5.10b).

The nucleic acids are very large structures composed of aggregates of these nucleotides. The two purines, adenine and guanine, are present in both DNA and RNA. The pyrimidines commonly found in RNA are uracil and cytosine; in DNA, thymine and cytosine. Nucleic acids are pictured structurally as a double helix (Watson, 1965). The DNA helical structure is illustrated in Fig. 5.10. Compare this structure to that of the protein α-helix structure (Fig. 5.2). An informative history of the research which led to the establishment of the α-helix structure for proteins and the double helix for DNA and to an understanding of biological structures is summarized by Olby (1974).

An almost endless variety of different nucleic acids is possible through variation of the nucleotide sequence, and specific differences among them are believed to be of the highest importance, for the nucleic acids are the main constituents of the genes, the bearers of hereditary information.

VII. PYRROLES AND PORPHYRINS

In addition to the carotenoids, a class of pigment molecules that are vitally important for life are the chlorophylls in the chloroplasts of plant

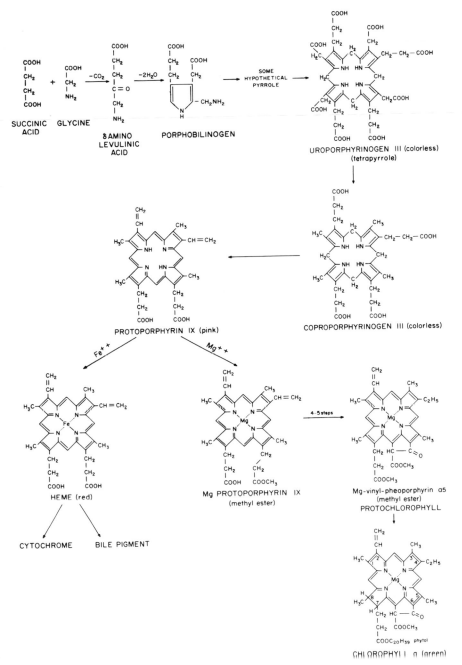

Fig. 5.12 Biosynthesis of porphyrins. This scheme for the biosynthesis of porphyrins, chlorophyll, and heme was developed by Granick (1950, 1958) and Shemin (1955, 1956).

(V)

cells for photosynthesis, hemoglobins of red blood cells, and cytochromes, the respiratory enzymes of the cell. These pigment molecules have a common structure, which consists of four pyrrole rings (V) linked by methine bridges (—CH=), forming a cyclic tetrapyrrole structure, named porphyrins. The iron porphyrins, the hemes, cytochromes, and other metal porphyrins such as the magnesium porphyrin chlorophyll are essential molecules for sustaining life on earth. The porphyrins are catalysts which serve ideally for electron transfer in photo and chemical reactions. The biosynthetic pathways for the synthesis of porphyrins, chlorophyll, and heme pigments and their structure is illustrated in Fig. 5.12. It has been observed that porphyrins in aqueous media will form liquid crystalline structures. For example, hemoglobin with water forms a crystalline structure in which the disc-shaped hemoglobin molecules pack into layers.

VIII. ADP–ATP SYSTEM

Another striking molecular feature that all living organisms have in common is the presence of an adenosine diphosphate (ADP) and adenosine triphosphate (ATP) system as the energy storing mechanism. This involves the synthesis of ATP from ADP and inorganic phosphate when surplus energy is being released, and the breakdown of ATP to ADP and phosphate when energy is required. The chemical energy of ATP is used to perform the chemical and mechanical work of the cell. The universal occurrence of the ADP–ATP system strongly suggests that it is the archetypical energy storing mechanism.

IX. REMARKS

These molecules, as we have illustrated, possess a structure of their own—they have definite shapes, sizes, and dimensions. Therefore, we have to see how at the molecular level we can relate their molecular structure to cellular structures, which are composed of these molecules.

An important question then is, could these molecules which we associ-

TABLE 5.1

**Molecules in Living Structures Claimed to Elicit
Liquid Crystalline Properties**[a]

Lipids
Lecithin
Sphingomyelin
Cephalin
Monoglycerides
Cholesterol esters
Various phospholipids
Proteins and polypeptides
Myosin
Hemoglobin
Trypsin
Poly-γ-benzyl-L-glutamate
Poly-γ-methyl-L-glutamate
Poly-γ-ethyl-D-glutamate
Poly-β-benzyl-L-aspartate
Poly-α-L-glutamic acid
Poly-α-sodium-L-glutamate
Poly-L-lysine hydrochloride
Nucleic acids
Deoxyribonucleic acid (DNA)
Ribonucleic acid (RNA)
Polysaccharides
Chitin

[a] Taken in part from Mishra (1975).

ate with life have interacted and, with water, formed liquid crystalline cellular structures? The helical structure illustrated for proteins and DNA form rods that pack into hexagonal geometries to give what we have described as a liquid crystalline structure. In fact, many proteins, nucleic acids, hemes, lipids, and polysaccharides in water exhibit liquid crystalline structures (Table 5.1). How such molecules formed into membranes and cellular structures is in the realm of speculation and is worth exploring in more detail.

REFERENCES

Chapman, D. (1973). Some recent studies of lipids, lipid-cholesterol and membrane systems. *Biol. Membr.* **2**, 91.
Etkin, W. (1973). Structure of the DNA molecule. *Bioscience* **23**, 653.

Granick, S. (1950). Magnesium vinyl pheoporphyrin a_5, another intermediate in the biological synthesis of chlorophyll. *J. Biol. Chem.* **183**, 713.

Granick, S. (1958). Porphyrin biosynthesis in erythrocytes. I. Formation of ξ-amino-levulinic acid in erythrocytes. *J. Biol. Chem.* **232**, 1101.

Mishra, R. K. (1975). Occurrence, fluctuations and significance of liquid crystallinity in living systems. *Mol. Cryst. Liq. Cryst.* **29**, 201.

Olby, R. (1974). "The Path to the Double Helix." Univ. of Washington Press, Seattle.

Pauling, L. (1960). "The Nature of the Chemical Bond," 3rd ed. pp. 498– 502. Cornell Univ. Press, Ithaca, New York.

Robinson, C. (1956). Liquid-crystalline structures in solutions of a polypeptide. Part I. *Trans. Faraday Soc.* **52**, Part 4, 571.

Robinson, C. (1958). Liquid-crystalline structures in solutions of a polypeptide. Part II. *Discuss. Faraday Soc.*, pp. 29–42.

Robinson, C. (1966). The cholesteric phase in polypeptide solutions and biological structures. *Mol. Cryst. Liq. Cryst.* **1**, 467.

Shemin, D. (1955). The succinate–glycine cycle: The role of δ-amino-levulinic acid in porphyrin synthesis. *Porphyrin Biosyn, Metab., Ciba Found. Symp., 1955* p. 4.

Shemin, D. (1956). The biosynthesis of porphyrins. *Harvey Lect.* **50**, 258.

Watson, J. D. (1965). "Molecular Biology of the Gene." Benjamin, New York.

Chapter 6
Molecules, Macromolecules, and Self-Organizing Systems

I. INTRODUCTION

Before we try to draw analogies from the structure of the molecules we have just described and their relationship to the structures of living cells, it is of interest to hypothesize how these molecules were organized into macromolecules and then into a self-organizing system, hence into a cell. We can envision the role of liquid crystals in macromolecular structures and in the organization of a cell.

Historically it is of interest to note that Pasteur, in 1860, shortly after Darwin's publication of "The Origin of Species" in 1859, showed by an ingenious set of experiments that life came from existing life on earth. This dispelled for a time belief in the spontaneous generation of life. However, Pasteur did not rule out the spontaneous generation of present day life, for in 1878 he wrote: "I have been looking for it (spontaneous generation) for 20 years but I have not yet found it, although I do not think it is an impossibility." In this same context Darwin, in a letter to Joseph Hooker in 1871, wrote:

> It is often said that all the conditions for the first production of living organisms are now present, which could ever have been present. But if (and—oh, what a big if) we could conceive in some warm little pond, with all sorts of ammonia and phosphoric salts, light, heat, electricity, etc., present, that a protein compound was chemically formed ready to undergo still more complex changes . . .

So we see that Darwin and Pasteur dwelled on the possibility of the spontaneous generation of life.

A new impetus for seeking the origins of life began to emerge in the late 1930s, beginning with the writing of Bernal (1951, 1967), Haldane (1928, 1966), Oparin (1938, 1968), and Urey (1952). Though the question as to the origin of life remains at present unanswered, let us briefly examine the various bits of experimental data suggestive of a physical–chemical basis for the origin of life. The philosophy on which this is based is that the environment was such that organic molecules were formed. This led to the generation of macromolecules, such as proteins, nucleic acids, and polysaccharides. These molecules then organized themselves into macromolecular filamentous structures and membranes. It is estimated that from such beginnings to the origin of life took about 3 billion years. Once organized, as in a cell, they were then capable of growth and replication.

II. SYNTHESIS OF ORGANIC COMPOUNDS

According to Oparin (1938, 1968), the primitive atmosphere was reducing and consisted of hydrogen, methane, ammonia, nitrogen, water vapor, and that the first carbon compounds were hydrocarbons. Haldane (1928, 1954, 1966) suggested that before the origin of life, organic compounds must have accumulated until primitive oceans reached the consistency of "hot diluted soup." He based this on the assumption that the primitive atmosphere contained CO_2, NH_3, and water vapor, but no oxygen. Haldane claimed that such a mixture exposed to ultraviolet radiation would give rise to a vast variety of organic compounds.

In looking for ways to synthesize complex organic molecules from simple molecules in a prebiological environment, Calvin (1969, 1975) irradiated carbon dioxide and hydrogen with high energy ionizing radiation in the cyclotron. Formaldehyde, formic acid, acetic acid, and other reduced carbon compounds were obtained. These molecules could then be used for further synthesis and a variety of complex organic compounds could be formed.

Important biological molecules that act as catalysts in chemical reaction are the porphyrins (Fig. 5.12). Calvin (1969) assumed that, by a process of autocatalysis, each of the steps in their synthesis is catalyzed by an iron-containing compound with the end product being protoporphyrin IX (Fig. 5.12). Since the end product can catalyze earlier steps in its own formation, the whole reaction series is regarded as autocatalytic. Once protoporphyrin IX was formed by a slow and random series of reactions, it served to increase the probability that the earlier steps would continue to occur. That this process, or something similar to it, became incorporated into living cells relatively early in evolutionary history is suggested by the universality of pyrrole in the organic world. The synthesis of porphyrins from some pyrrole and the biosynthesis of chlorophyll, heme, and cytochrome from protoporphyrin IX is illustrated in Fig. 5.12.

Miller (1953, 1955, 1957) was one of the first to synthesize amino acids in a prebiological simulation experiment. He applied electric discharges to a mixture of methane, ammonia, hydrogen, and water vapor. The amino acids he identified were glycine, alanine, β-alanine, aspartic acid, α-amino-n-butyric acid (Fig. 5.1), and other organic compounds, but purines and pyrimidines were absent. When similar gaseous mixtures were subject to high temperatures (30°–90°C), ultraviolet radiation, ionizing radiation, and electrical discharge, analysis of the contents of these reaction chambers showed that small quantities of a great variety of organic molecules had been produced. These molecules are of considerable biological importance, for they include many organic acids of low molecular weight—amino acids, including glycine, alanine, and adenine, urea, and simple sugars such as ribose. Therefore, it is not unreasonable to find that a variety of organic molecules found in living systems can be synthesized in the laboratory without the agency of living cell.

A. Optically Active Molecules

How optically active compounds formed has challenged science since Pasteur's early experiments. These experiments with crystals of tartaric acid and sodium ammonium tartrate isomers (racemic forms) led Pasteur to an important observation—that asymmetric molecules are always the products of living processes. This appeared to Pasteur to be one of the fundamental differences between the chemistry of life and that of inanimate matter. He postulated that this peculiar asymmetry might be the manifestation of asymmetric forces in the environment acting at the time of molecular synthesis on the evolving cells.

Studies of liquid crystals may be helpful in looking at the asymmetry of biological molecules. A molecule that has an asymmetric atom can convert a nematic liquid crystal to a cholesteric–nematic liquid crystal. This structure will rotate polarized light. A mechanically twisted nematic liquid crystal cell can also be prepared in the laboratory by rubbing two glass surfaces in the same direction and then placing them so that the direction of rubbing of one is perpendicular to the direction of rubbing of the other. A nematic liquid crystal placed between these two glass plates will rotate polarized light through 90°.

We can speculate that a similar situation in the evolutionary process may have occurred, for if organic compounds such as p-azoxyanisole were synthesized, a nematic liquid crystal was already in the environment. If a solid surface (inorganic or organic) were exposed to environmental forces it could develop charges on the surface. The charges on the surface would interact with the polar groups in the liquid crystal molecule, thus orienting the liquid crystal. The surface charges could come from ions in the environment, from directed abrasion on the surfaces, from the action of water, and from wind moving in a given direction. These various forces acting on the surface will align the charges and as a result will orient the liquid crystal molecules so that their long axes are parallel. If another surface is similarly acted on and these two surfaces make contact so that one is perpendicular to the other, a twisted nematic liquid crystal can be formed. The resulting twisted nematic liquid crystal can be used as a solvent for carrying on organic reactions that result in the preparation of dextro or levo forms or a racemic mixture. Thus natural forces could have synthesized an optically active compound in a twisted nematic liquid crystal (to give the dextro or levo form, depending on whether the twist of the liquid crystal solvent has a left-handed or a right-handed twist). Such an optically active molecule would then have served as an intermediate for the synthesis of other cholesteric–nematic-type liquid crystals. That such a process may have taken place in the synthesis of polypeptides and proteins is worth further consideration. At this point we can say that physical and chemical forces shaped the molecules in particular molecular geometries which became incorporated in the living cell. Once the molecules became organized into specific structures, e.g., cell membranes, and became functional as a cell, they remained similarly organized in cells which evolved.

III. STRUCTURAL MODELS

Claude Bernard (1866, 1878) expressed the idea that matter, including organic matter, is inherently inert, but organized matter as structured in a cell is able to respond to external stimuli. This response to external stimuli is repeatedly asserted by Bernard to be the fundamental characteristic of life.

How did a self-organizing system get started and what kind of models can we prepare that would satisfy the requirements for a cell? Bernal (1951) suggested that clays would provide sufficient surfaces for organic molecules to be adsorbed, and in the presence of a catalyst, the synthesis of macromolecules could have taken place rapidly. For example, vinyl monomers, either adsorbed on the external surfaces or between the lamellae of clay, have been shown to polymerize. The view is that organic material separated at different phase boundary layers in the sediment strata during certain temperature stages in the development of the earth's surface is an attractive one, for found among these layers were organic materials in agglomerates containing phospholipids which could have served as boundary membrane layers. Radiation, temperature variations, and diurnal alterations on the earth's surface became important factors at

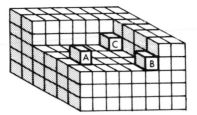

Fig. 6.1 Depiction of crystal growth on a crystal face. A is held by one face, B has two contact surfaces, and C has three. The crystal therefore grows through a progressive addition of molecules.

these boundaries which provided special conditions for the synthesis of organic molecules, their polymerization, and aggregation into macromolecular structures.

Crystallites may also have served as a template for a self-replicating system, for the process of crystal growth—that is, the progressive accretion of molecules on a crystal surface—could have occurred (Figs. 6.1 and 6.2). Cairns-Smith (1971) has developed such a model as a possible mechanism for the growth and organization of macromolecules into a cell.

There are several other model systems which should be mentioned. One is known as the Liesegang phenomenon. Liesegang (1907) observed, in staining nerve tissue by means of the Golgi technique for histological study, that the staining occurred in a periodic manner. These experiments are described and discussed in great detail by Hedges (1932) and by Stern (1954). The Liesegang phenomenon can be observed if a drop of 15% silver nitrate is placed on gelatin which has previously been impregnated with about 0.4% potassium dichromate. The silver slowly diffuses into the gelatin, there reacts with the potassium dichromate, and a colored silver dichromate complex is formed. The formation of the complex occurs in waves through the gelatin, which results in a series of concentric rings, phase boundary separations, in the clear gel (Fig. 6.3). An analogous system is that of certain dye-complex salts in colloids or proteins, e.g., potassium dichromate in gelatin or in serum. It will be observed that phase boundaries begin to form around the periphery of the drop and proceed by periods of rapid and slow growth similar to that illustrated in Fig. 6.3. Light can modify these periodic structures if the complex formed is light-sensitive. In both examples they appear to pass through a liquid crystalline phase.

Another is periodic wave phenomena in solution; this has more recently been studied by Winfree (1972, 1973). He modified a reagent originally prepared by Zaikin and Zhabotinsky (1970). The mixture was dispersed in

Fig. 6.2 A section through a biosynthesized crystal found in the fungus *Phyco-myces blakesleeanus.* Note lattice structure. From Wolken (1971).

water, but it is convenient to gel the reagents with an equal volume of chromatographic silicon dioxide. The reagent mixture consists of sodium bromate, sodium bromide, malonic acid, and phenanthroline iron(II) sulfate. The mixture is poured into a petri dish and illuminated from below. What is observed is propagating waves, with orange regions [iron(II) phenanthroline] and light blue regions [iron(III) phenanthroline]. The structures formed resemble a liquid crystalline structure (Fig. 6.4). This is the result of chemical activity (oxidation–reduction reactions) propagated through the liquid. The metal ion (or ion complex) catalyzes the oxidative decarboxylation acid by bromate in an aqueous solution. Winfree has

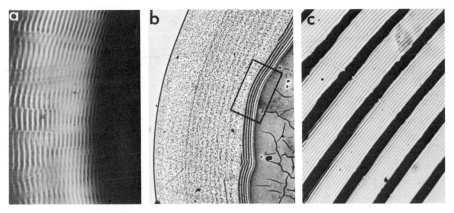

Fig. 6.3 Liesegang phenomena, as seen through the light microscope, when a drop of silver nitrate is placed on a film of gelatin saturated with potassium dichromate.

likened such oxidation–reduction phenomena to action potentials in a nerve membrane through the excitable medium.

IV. CELLULAR MODELS

Let us turn to two models that resemble a cell more closely in appearance. These studies go back to the experiments of Leduc (1911), who formed structures from a variety of inorganic salts and organic compounds that resembled cells. Such experiments were appealing and influenced many attempts to create models for a living cell. Budenberg de Jong (1936) studied colloidal systems and was interested in building high molecular weight compounds. Oparin (1938) was impressed by these studies in his search to find a dynamic colloidal system as a model for cellular behavior. One such model system was originally made by mixing gelatin with gum arabic in water to 42°C until a clear solution was obtained; others were made with gelatin and lecithin and a variety of other substances. The important observation Oparin made was that at a critical pH, droplets came out of solution; he named these "coacervates." The fact that they form into droplets or micropheres rather than in layers convinced him that coacervates could function as an experimental model for a cell, and could further his origin of life studies.

A series of experiments based on ideas about the prebiological environment was begun by Fox in the early 1960s to bring about the synthesis of complex organic compounds from amino acids. Initially, the method con-

Fig. 6.4a,b Zaikin and Zhabotinsky reagent, modified by Winfree (1973), showing the waves of chemical activity propated through the liquid at room temperature. These waves are seen as oxidation–reduction potentials of metal ions present in catalytic amounts. The metal ion (or ion complex) catalyzes the oxidative decarboxylation of an aliphatic acid by bromate in an acidic aqueous solution.

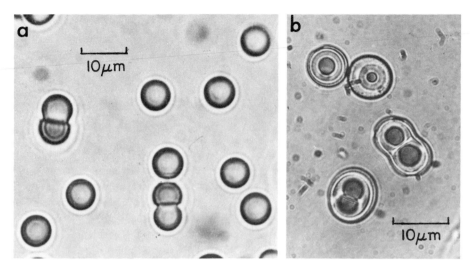

Fig. 6.5 Proteinoid microspheres, prepared by the polymerization of amino acids according to Fox (1965). (Courtesy of S. Fox, Institute of Molecular Evolution, University of Miami, Miami, Florida.)

sisted of heating a mixture of amino acids to temperatures of 160°–200°C for several hours under anhydrous conditions in an atmosphere of nitrogen. The mixture contained aspartic acid, glutamic acid, and lysine (Fox *et al.*, 1963). Using thermal polycondensation they were able to copolymerize amino acids, which they named *proteinoids*. On analysis, 18 amino acids found in proteins were identified. The molecular weights of these proteinoids ranged from 3000 to 9000 depending on the method of preparation. Interestingly, when proteinoid is treated with hot water, microscopheres separate out of the cooling clear solution (Fig. 6.5).

The proteinoid microspheres (Fig. 6.5) can vary in size from a few microns to 100 μm in diameter, are relatively stable, and, depending on their preparation, exhibit living-cell-like behavior. That is, they possess a cell membrane, and they bud, coalesce, increase in size, and then divide, a process analogous to the growth of bacterial cells.

Coacervates and proteinoid microspheres are a long way from being the living state. If left alone, such systems come to an equilibrium with no exchange of energy. Therefore, it is necessary to incorporate in these microspheres the right kinds of molecules to provide coordinated reversible (oxidation–reduction) reactions whose response to environmental changes will produce the internal adjustments necessary to stabilize the whole structure. Photosensitive dyes, chlorophyll, enzymes, and nucleic acids have been incorporated into these microspheres. Such experiments have stabilized, to some extent, these cellular model systems and resulted in activities which simulate many of the behavioral chemistry of living cells.

A. Molecular Self-Ordering

Liquid crystals provide an interesting model for a self-replicating system. For example, poly-γ-benzyl-L-glutamate in chloroform forms a large number of spherulites upon evaporation (Robinson, 1956). The spherulites coalesce and result in a birefringent material that spontaneously orients itself at the air interface. The birefringent structure shows microscopic visible periodicities and indicates a helical arrangement. These spherulites (Fig. 6.6), like proteinoids (Fig. 6.5), are enclosed with a membrane and divide by passing through an oblate spheroid phase. This "self-orientation" exhibited by spherulites provides a model then for how molecules form oriented structures that are of biological importance (Filas, 1977; King, 1969; Robinson, 1966).

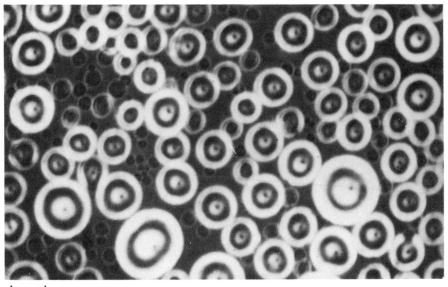

10 μm

Fig. 6.6 Spherulites near the transition temperature between nematic liquid crystals and the isotropic liquid. Note that the spherulites are in a dynamic state of transition. The small spherulites coalesce with the larger ones, which can then divide.

Many molecules, particularly elongated molecules in solution, will self-organize and orient. The orientation is such that the long axes of the molecules align parallel to one another. This alignment can take place in the liquid or liquid crystalline state. Liquid crystals which possess one- and two-dimensional order will self-organize and orient. Molecular ordering of molecules in liquids was recognized in the late 1920s, but too little experimental work was done at the time to unravel the structure of these liquids.

Brady (1973, 1974) more recently studied the molecular aggregation of alkyl monoiodides and monobromides in the solvent Decalin (decahydronaphthalene). Brady (1974) found that with C_{20} alkane chains, the molecules arrange themselves parallel to one another and form a spherical aggregate whose radius is 17 Å. This radius length is equivalent to the measured length of the dissolved chain with the head groups making up the periphery. In this model the molecules form an aggregate in which their end groups are arrayed on the surface of a sphere (refer to Chapter 3). If chains have about the same lengths, the diameter of the sphere is determined by the length of the longer molecules. Shorter molecules interweave with the longer ones so that their end groups are also on the periphery. Addition of extra links to the chains increases the attraction, not only because of greater flexibility brought about by the possibility for rotation about the C–C bonds.

The morphology depends on the concentration of the solute aggregates and indicates that there is a cluster of spherical micelles in dilute solutions. In more concentrated systems the spherical aggregates are viscous and exhibit isotropic liquid crystalline states which pack in a cubic ar-

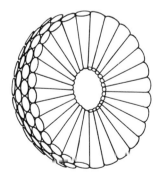

Fig. 6.7 A schematic representation of the aggregate structure of a cluster of long-chain molecules such as found in an alkyl halide. The pattern formed by the end groups is interesting, considering that only alkyl chains with polarizable head groups are involved.

rangement. With alkyl iodides the iodide ends of the long chains make up the outer surface of the sphere. Figure 6.7 is a schematic representation of a self-ordered and oriented long-chain alkyl halide. Other self-organized molecular aggregates commonly found in nature organize into lamellar and hexagonal structures (Chapter 3).

The aggregation of chain molecules is important in the chemistry of liquid crystalline structures as in the formation of micelles (soaps and detergents), but more important, it may be a mechanism in the structural formation of cellular membranes within living organisms.

V. REMARKS

A variety of organic molecules found in living cells are synthesized in the environment and many have now been synthesized in the laboratory. The question then is, how did these molecules assemble into membranes and then into organized structures, resulting in a self-reproducing cell? As to the organization of living matter, Tanford (1973, 1978) has examined this question in more detail, as have Cairns-Smith (1971), King (1969), and Morowitz (1967).

However, the search for self-ordering and replicating structures has suggested from the studies of Robinson (1966) that liquid crystals provide a model system for investigating self-assembly and replicating structures. It is this point that we have tried to emphasize. But, if we are to delve into how liquid crystalline systems may underlie cellular structures, and hence into a functional cellular system, it is necessary to turn to the structural organization of living cells.

REFERENCES

Bernal, J. D. (1951). "The Physical Basis of Life." Routledge & Kegan Paul, London.
Bernal, J. D. (1967). "The Origin of Life." World Publ. Co., Cleveland, Ohio.
Bernard, C. (1866). "Leçons sur les propriétés des tissus vivants." Baillière et Fils, Paris.
Bernard, C. (1878). "Leçons sur les phénomènes de la Vie Communs aux animaux et aux végétaux." Baillière et Fils, Paris.
Brady, G. W. (1973). Effect of length on the interaction of dissolved long chain molecules. *J. Chem. Phys.* **58**, 3542.
Brady, G. W. (1974). On the aggregation of dissolved alkane chain molecules. *Acc. Chem. Res.* **7**, 174.
Budenberg de Jong, H. (1936). "The Coacervation." Hermann, Paris.
Cairns-Smith, A. G. (1971). "The Life Puzzle." Oliver & Boyd, Edinburgh.
Calvin, M. (1969). "Chemical Evolution." Oxford Univ. Press, London and New York.
Calvin, M. (1975). Chemical evolution. *Am. Sci.* **63**, 169.

Darwin, C. (1859). "On the Origin of Species by Means of Natural Selection, or, the Preservation of Favored Races in the Struggle for Life." Murray London (republished by Modern Library, New York, 1936).

Darwin, C. (1892). Letter to Joseph Hooker, 1871, in "The Autobiography of Charles Darwin and Selected Letters" (F. Darwin, ed.), p.220. Appleton, New York (republished by Dover, New York, 1958).

Filas, R. J. (1977). Tactoidal shell defects in poly-γ-benzyl-D-glutamate) liquid crystals. J. Phys. 39, 49.

Fox, S. W. (1965). A theory of macromolecular and cellular origins. Nature (London) 205, 328.

Fox, S. W., and Dose, K. (1972). "Molecular Evolution and the Origin of Life." Freeman, San Francisco, California.

Fox, S. W., and Harada, K., Woods, K. R., and Windsor, C. R. (1963). Amino acid compositions of proteinoids. Arch. Biochem. Biophys. 102, 439.

Haldane, J. B. S. (1928). "Possible World." Harper, New York.

Haldane, J. B. S. (1954). "The Biochemistry of Genetics." Macmillan, New York.

Haldane, J. B. S. (1966). "The Causes of Evolution." Cornell Univ. Press, Ithaca, New York.

Hedges, E. J. (1932). "Liesegang Ring and Other Periodic Structures." Chapman and Hall, London.

King, L. J. (1969). Biocrystallinity. Bioscience 19, 505.

Leduc, S. (1911). "Mechanisms of Life." London.

Liesegang, R. E. (1907). The formation of layers during diffusion. Z. Phys. Chem. 59, 444.

Miller, S. L. (1953). A production of amino acids under possible primitive earth conditions. Science 117, 528.

Miller, S. L. (1955). Production of some organic compounds under possible primitive earth conditions. J. Am. Chem. Soc. 77, 2351.

Miller, S. L. (1957). The mechanism of synthesis of amino acids by electric discharges. Biochim. Biophys. Acta 23, 480.

Morowitz, H. J. (1967). Biological self-replicating systems. Progr. Theoret. Biol. 1, 38.

Oparin, H. I. (1938). "The Origin of Life" (S. Morgulis, transl.). Macmillan, New York (2nd ed., Dover, New York, 1953).

Oparin, H. I. (1968). "Genesis and Evolutionary Development of Life." Academic Press, New York.

Pasteur, L. (1860). De l'origine des ferments. Nouvelle expériences relatives aux générations dites spontanées. C. R. Hebd. Scances Acad. Sci. 50, 849.

Pasteur, L. (1878). In "Collected Works of Pasteur (Oeuvres de Pasteur)," by Vallery-Rodot (1922–1939), Vols. I and II. Masson, Paris.

Robinson, C. (1956). Liquid-crystalline structures in solution of a polypeptide. Trans. Faraday Soc. 52, 571.

Robinson, C. (1966). The cholesteric phase in polypeptides and biological structures. Mol. Cryst. 1, 467.

Stern, K. H. (1954). The liesegang phenomenon. Chem. Rev. 54, 79.

Tanford, C. (1973). "The Hydrophobic Effect: Formation of Micelles and Biological Membranes." Wiley, New York.

Tanford, C. (1978). The hydrophobic effect and the organization of living matter. Science 200, 1012.

Urey, H. (1952). "The Planets." Yale Univ. Press, New Haven, Connecticut.

Winfree, A. (1972). Spiral waves. Science 175, 634.

Winfree, A. (1973). Scroll shaped waves of chemical activity in three dimensions. *Science*
    181, 937.
Winfree, A. T. (1977). Spatial and temporal organization in the Zhabotinsky Reaction. *Ado.
    Biol. Med. Phys.* **16**, 115.
Wolken, J. J. (1971). "Invertebrate Photoreceptors: A Comparative Analysis, p. 147.
    Academic Press, New York.
Zaikin, A., and Zhabotinsky, A. (1970). Concentration wave propagation in two-dimensional
    liquid-phase self-oscillating system. *Nature (London)* **225**, 535.

Chapter 7
The Cell and Cellular Structures

> . . . living systems actually are liquid crystals or it would be more correct to say, the *paracrystalline* state undoubtedly exists in living cells.
>
> *J. Needham (1950)*

I. INTRODUCTION

If we believe in evolution, then every form of life shares a common ancestry, and all forms of life as we now see it arose from a cell. Various cells aggregated for mutual benefit and specialized to produce the whole integrated plant or animal. Robert Hooke in 1665, using a simple microscope, is credited with the recognition of the cell as the unit of life and the fact that a cell cannot exist without a cell membrane. The elucidation of the structure of the cell has gone hand in hand with the developments in optics. The development of the compound microscope in the 1830s and its application to biological tissue made it possible to begin to resolve cells. From these beginnings, Rudolf Virchow (1858) recognized that each cell originates from another cell. This discovery marked the beginning of the science of cellular systems.

Further developments in optics for microscopy, such as phase, polarization, interference, and fluorescence, have revealed considerable structural information about the cell. With the introduction of electron microscopy in the early 1940s and its continued application to biology, the cell has been shown in its structural complexity hitherto unknown. Therefore, electron microscopy (including scanning electron microscopy) together with optical diffraction, X-ray diffraction, spectroscopy, and other analytical tools, are now bringing us closer to a molecular description of cellular structures.

The living cell is a complex dynamic system enclosed within a membrane, concerned with the processes of energy—processes vital to its maintenance, growth, and reproduction.

Organisms are separated into primitive forms and advanced forms. Bacteria and blue-green algae are considered primitive, for they do not possess a membrane-enclosed nucleus and have their genetic material dispersed throughout their cytoplasm. Such organisms are classified as *prokaryotes* (Fig. 7.1). Cells found in the more highly evolved or advanced forms of organisms are *eukaryotes,* which have a membrane-enclosed nucleus, mitochondria, and other membrane-bound organelles (Fig. 7.2).

Cells vary in shape and size and structural organization depending on their environment and their function. Primitive unicellular organisms exhibit diverse behavior associated with more highly evolved plant and animal cells, but they and all cells have certain structures and biochemistry in common. These unicellular organisms respond to light, gravity, magnetic fields, and the presence of nearby objects, as well as to change in their chemical environment. Why is all this sensory behavior built into such primitive organisms? Claude Bernard, in 1866, recognized that the response to external stimuli—*irritability*—is characteristic of life. He attributed this to the fact that the organic matter in the cell is very specially organized, and it is this organization that permits the cell to detect physical and chemical changes in its environment. Since these primitive cells can sense their environment, what can be learned from them about the development of sensory systems of more highly evolved plant and animal cells? As organisms, they possess no obvious structures that resemble a nervous system. What kind of macromolecular structures of membranes and receptors do they have that permit them to detect environmental changes and to act as an integrated organism? It is in search of this very special organization, their molecular structure for sensing their environment, that draws our attention.

One of the unicellular organisms, used here as a model to illustrate the

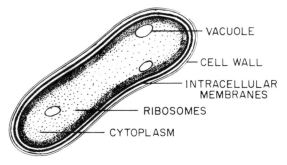

Fig. 7.1 Prokaryote. Schematized bacterial cell.

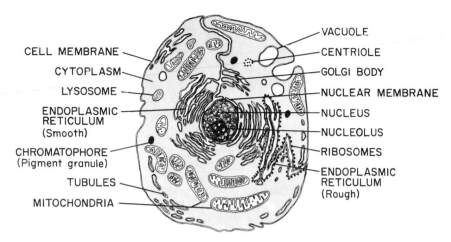

CELL MEMBRANE
CYTOPLASM
LYSOSOME
ENDOPLASMIC RETICULUM (Smooth)
CHROMATOPHORE (Pigment granule)
TUBULES
MITOCHONDRIA

VACUOLE
CENTRIOLE
GOLGI BODY
NUCLEAR MEMBRANE
NUCLEUS
NUCLEOLUS
RIBOSOMES
ENDOPLASMIC RETICULUM (Rough)

Fig. 7.2 Eukaryote. Schematized animal cell structure.

structure of the cells, is the protozoan algal flagellate, *Euglena gracilis* (Fig. 7.3). The structure of *Euglena* and its chemistry is profoundly altered, depending on whether it lives in light or in darkness (Wolken, 1967). When living in the light, it develops chloroplasts and photosynthesizes like a plant, but in continuous darkness, it loses its chloroplasts and hence its photosynthetic ability. As a result, it lives by chemosynthesis, a process typical of all animal cells. This ability to adapt to light or darkness is fully reversible, providing that mutations do not occur. The light (pho-

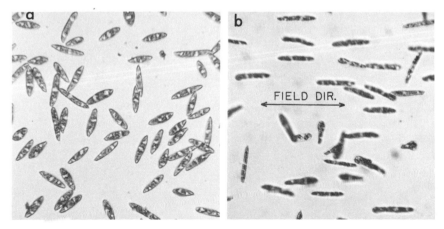

Fig. 7.3 *Euglena gracilis.* (a) Cells in the light swimming at random. (b) Cell oriented in a radio frequency field. From Wolken (1967), *"Euglena,"* p. 88 Reprinted by permission.

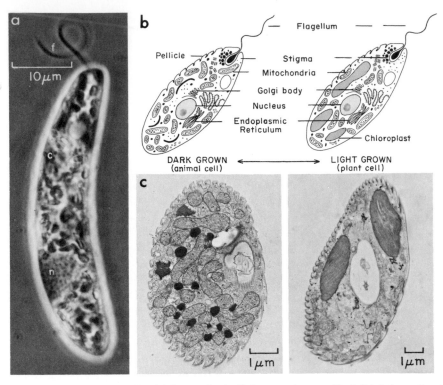

Fig. 7.4 *Euglena gracilis.* (a) Grown in the light, greatly magnified. (b) Schematized cell structure from growth in darkness to growth in light. (c) Cross-sections through the dark-grown and light-grown organisms. From Wolken (1971).

tosynthesis) ↔ dark (chemosynthesis) adaptation brings about changes in the chemistry and internal structure of the organism (Fig. 7.4). Mutations to nonphotosynthetic *Euglena* can be easily produced by changing the physical environment, for example, growth temperature near 37°C, ultraviolet and ionizing radiation, and high pressure.

The cellular structures observed in *Euglena* are typical of the organelles found in a variety of plant and animal cells such as chloroplasts, flagella, nuclei, mitochondria, endoplasmic reticulum, and Golgi (Figs. 7.4 and 7.5).

II. STRUCTURES

A. Exoskeleton

Certain cells have in addition to a cell membrane a protective coat, exoskeleton or pellicle, an outer cell wall, that surrounds the cell. The eugle-

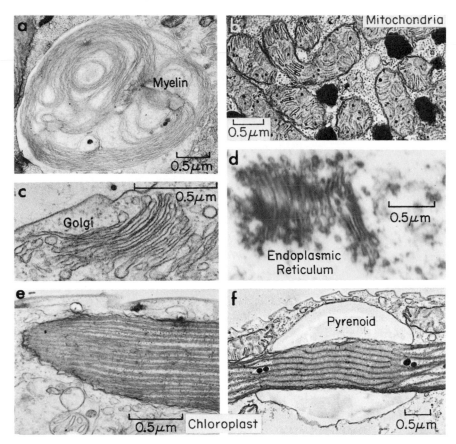

Fig. 7.5 Electron micrographs of organelle structures as seen in the *Euglena:* (a) myelin, (b) mitochondria, (c) Golgi, (d) endoplasmic reticulum, (e) chloroplast, (f) chloroplast in relation to the pyrenoid (center for starch synthesis). From Wolken (1975, p. 54).

noid exoskeleton, or pellicle, has been of interest to biologists for a long time because of the organism's characteristic cellular motion known as euglenoid movement.

The pellicle is a membranous structure which is so arranged and joined over its entire surface that it permits the organism to elongate and contract. This behavior suggests that the pellicle had similarities to that of muscle fibers. But chemical analysis so far has not identified musclelike proteins in the pellicle. However, proteins and lipids have been found to be associated with isolated pellicle material. The present evidence indicates that the major structural component is the polysaccharide polymer, chitin (see Fig. 5.11).

The pellicle striations are seen under light and electron microscopy as longitudinal helices (Fig. 7.6a,b). Electron microscopy of sectioned cells shows that the pellicle consists of interlocking fibers and membranes which pass helically along the cell, as a left-handed helix. The structure also reveals a system of semirigid rings that alternate with strips of soft pliable membrane (Fig. 7.6c). Examination of a large number of *Euglena* in various states of elongation and contraction gives the impression that structurally it is very much like that of a collapsible cup.

Such a helical structure of chitin fibers for the pellicle indicates similarities to the arthropod cuticle and to the ascidian tunic (*Ciona intestinalis*). An analogy of these structures to cholesteric liquid crystals will be brought out in more detail in Chapter 9 in relation to our discussion of the compound eye corneal lens of arthropods.

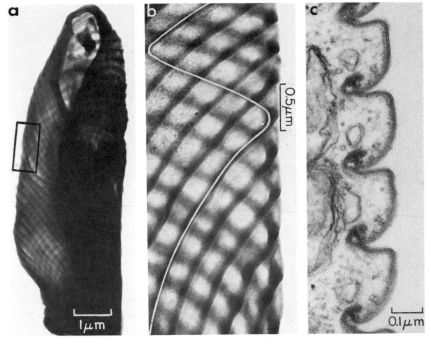

Fig. 7.6 Exoskeleton or pellicle of *Euglena*. (a) Ghost removed by sonication of cells, (b) enlarged area of (a) showing the helix pattern of the ridges of the pellicle, (c) electron micrograph section showing the double wall membrane of the ridges in the pellicle.

TABLE 7.1

Lipid Composition in Membranes[a,b]

	Animal cell membranes				Bacterial cell membranes	
	Myelin	Eryth-rocyte	Mito-chondria	Micro-some	Azoto-bacter agilis	Escher-ichia coli
Cholesterol	25	25	5	6	0	0
Phosphatidyl-ethanolamine	14	20	28	17	100	100
Phosphatidylserine	7	11	0	0	0	0
Phosphatidylcholine	11	23	48	64	0	0
Phosphatidylinositol	0	2	8	11	0	0
Sphingomyelin	6	18	0	0	0	0
Cerebroside	21	0	0	0	0	0

[a] Data taken in part from Korn (1966).
[b] Values given in percent.

B. The Cell Membrane

Let us begin with the structure of the cell membrane before discussing specific cellular membranes in more detail, since the cell membrane is so important to the integrity of the cell.

The cell membrane separates the external environment from the internal environment. In order to allow for the differential diffusion of ions and the exchange of gases, it has selective properties. The cell membrane was at one time envisaged as a passive barrier to diffusion and permeability. It is now thought to play an active role in chemical transport, energy transduction, and information transfer to and from the cell.

Biological membranes depend on the physical–chemical properties of lipids and proteins. The total lipids, the sterols and the phospholipids, are found in concentrations greater than 30% in a typical membrane (Table 7.1). The lipids have the unique property of forming mono- or bimolecular layers (see Fig. 5.6). This is due to the presence of the hydrophilic (water soluble) groups at one end of the molecule and hydrophobic (lipid soluble) groups at the other end of the molecule.

The cell membrane is described as a bilayer of lipids and proteins. The lipid layer consists of sterols, such as cholesterol and phospholipids. Every type of membrane has a unique set of proteins to account for its function. The cell membrane molecular structure as schematized by Danielli and Davson (1935) is referred to as the *unit membrane* (Fig. 7.7).

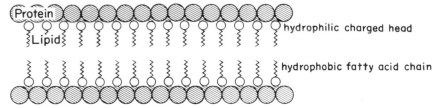

Fig. 7.7 Cell membrane. Schematic drawing of bilayer.

They assumed that a functional protein layer was absorbed over the lipid surfaces. Recent studies by electron microscopy and X-ray diffraction indicate that the cell membrane structure is a bilayer, of the order of 80 Å thick, with each layer of the bilayer being about 40 Å thick (Fig. 7.8). How the proteins and enzymes fit into the bilayer is not yet understood, but we will attempt to explain membrane structures in more detail in Chapter 11.

C. The Cytoplasm

The cytoplasm of cells consists of a variety of protein and lipid macromolecules. More then 20 to 30% of the cell's cytoplasm consists of proteins that can form filaments. In prokaryotes, there are no discrete membrane-bound structures (Fig. 7.1). In eukaryotes, structures, (organelles) are found that are membrane bound and have specific functions. It is possible to centrifuge a single cell, and by using the techniques of differential centrifugation, the cytoplasmic organelles and particles become stratified according to their densities. By this method, the organelles can be isolated from the cell and their biochemistry studied by various chemical and physical techniques. The molecular structure of these organelles and their chemistry are far from completely understood, but a great deal is known about them. They all appear to be systems of membranous structures.

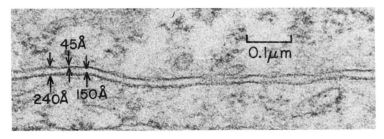

Fig. 7.8 Electron micrograph of cell membrane.

D. Nucleus

The most conspicuous body in eukaryotic cells is the nucleus, which is separated from the cell cytoplasm by a nuclear membrane (Fig. 7.9). The center of the nucleus is occupied by a large ovoid nucleolus. The nucleus may contain from a few to many nuclei and much granular material.

Cells which undergo nuclear division have a set of chromosomes which contain within their genes the hereditary molecules DNA and RNA (see Fig. 5.10). The chromosomes are most pronounced during nuclear division, a process known as mitosis (Fig. 7.10).

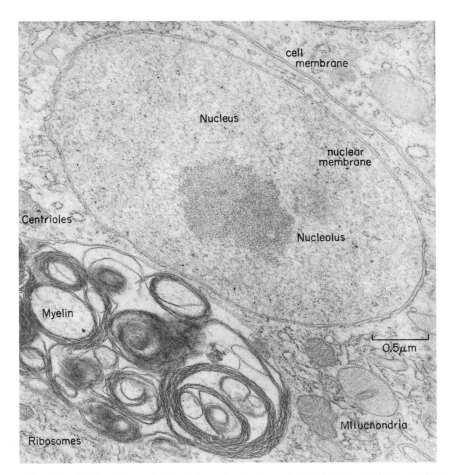

Fig. 7.9 Section through *Daphnia pulex*. Note cell membrane, nucleus, nuclear membrane, mitochondria, myelin, centrioles, and ribosomes in cytoplasm. Compare to Fig. 7.2.

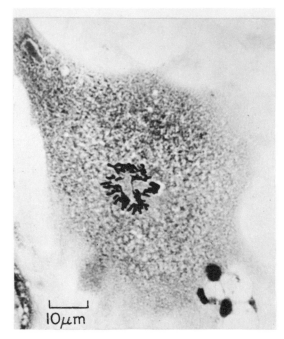

Fig. 7.10 Cell in tissue in culture undergoing nuclear division, mitosis. Note separating of chromosomes. From Wolken (1966), "Vision: Biochemistry and Biophysics of the Retinal Photoreceptors." Reprinted by permission.

E. Mitochondria

The mitochondria are the most numerous of the cytoplasmic organelles. They contain the respiratory enzymes for the cell. Mitochondria are not particularly different in structure from those found in either plant or animal cells (Fig. 7.11). Lipids, mainly phospholipids, make up 40 to 50% of the dry weight of mitochondria. Of these, chemical analysis shows that phosphatidylethanolamine, phosphatidylcholine, and cholesterol are the major phospholipids (Table 7.1).

Structural studies by electron microscopy indicate that mitochondria possess two membranes, a limiting membrane and an inner membrane. A system of ridges protrudes from the inside surface of the inner membrane. These ridges have been designated as *cristae*. It has been suggested that the oxidative enzymes of the mitochondria are built into these cristae. The inner membrane consists of an assembly of closely packed repeating units. These repeating units of the membrane are the basic transducers. The primary transduction of the mitochondrion is oxidative phosphorylation (Lehninger, 1965). In the process, electrons move along an electron

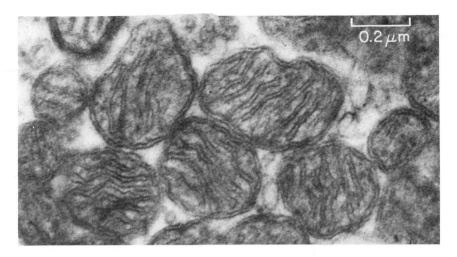

Fig. 7.11 Mitochondria, in inner segment of retinal rod cell of frog. From Wolken (1966), "Vision: Biochemistry and Biophysics of Photoreceptors." Reprinted by permission.

transfer chain and are linked to the synthesis of adenosine triphosphate (ATP). The structure of the mitochondrion is schematized in Fig. 7.12. It is best represented by a bilayer of lipid and protein.

An interesting relationship in the mitochondrion structure was pointed out by Green and Young (1971). The inner mitochondrion membranes undergo phase transitions from the lamellar state to a twisted state, which can be related to whether the mitochondrion is in an energized or nonenergized state. Their experimental studies provide support that the energized-twisted conformation is actually in the pathway of oxidative phosphorylation. There is evidence in the liquid crystalline structure for the transformation from the lamellar to a twisted structure. But whether such structural changes can be related to a liquid crystalline state and to their functional biochemistry needs further investigation.

F. Endoplasmic Reticulum and Golgi

The endoplasmic reticulum is an organization of membranes (Figs. 7.2, 7.5d, and 7.13). This complex cytoplasmic system of vesicles and tubules was first revealed by electron microscopy, which showed that its elements may be integrated in a continuous network. It is the system for protein synthesis to which ribosomes for RNA synthesis are associated. Another cluster of membranes with numerous vesicles is called the *Golgi body* or apparatus (Fig. 7.5c).

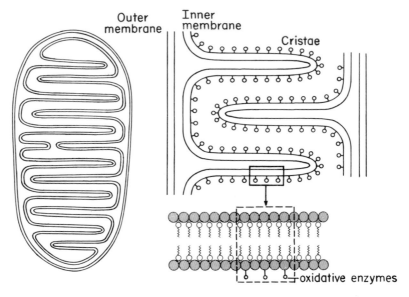

Fig. 7.12 Structural model of mitochondrion.

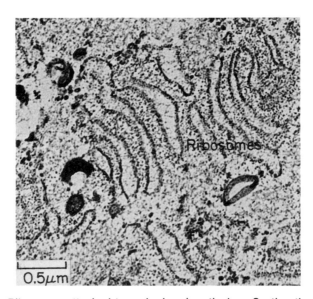

Fig. 7.13 Ribosomes attached to endoplasmic reticulum. Section through bovine cell in tissue culture.

G. Myelin

There are a variety of membranous structures found in cells. One of these is described as myelin figures which consist of concentric layers (membranes) of phospholipids (Figs. 7.5a and 7.9). Such myelin figures have been characterized as liquid crystal structures (Stewart, 1974). Phospholipids forming myelin figures (Fig. 5.7), as observed in the cells, have been used as models for studies of the structure of cell membranes. Myelinated nerves have a sheath which surrounds the axon of the nerve fiber (Figs. 10.4–10.7). Here, the myelin sheath is formed by bimolecular layers of lipid and protein.

H. Cilia and Flagella

Cilia and flagella are primitive motile organelles. Their motion propels the cell through its medium. Or, if the cell is fixed in place, the cilia move things past it. In the tissue cells of higher animals, some flagella and cilia have been modified to serve sensory functions.

The flagellum of *Euglena* is its effector for locomotion (Fig. 7.4). It is of the order of 30 μm in length and about 0.3 μm in diameter. Attached to the flagellum are lashlike cilia, referred to as *mastigonemata* and the flagellum has numerous segments along its entire length. The flagellum consists of a number of elementary filaments (axonemata) embedded in a matrix and covered by a membrane. There are eleven elementary filaments, of which nine (paired microtubules) are peripherally located while the other two are sometimes found in the center of the flagellum (Figs. 7.14 and 7.15). The pattern is known as the "9 + 2 array," and it is associated with all motile cells. This arrangement is observed in a variety of plant and animal cells with flagella or cilia, from bacteria to the sperm tails of man.

Microtubules from any kind of eukaryotic flagella and cilia are composed of a protein called *tubulin*. Among prokaryotes, the flagella or cilia are simpler. They are small and single stranded, consisting of a closely related protein, *flagellin*. Electron microscopy of bacterial flagella shows that they are constructed of globular subunits and that the subunits are arranged in helices of various kinds, depending on the type of bacterium studied. The bacterial flagellin protein shows, from X-ray diffraction and other studies, an α-helix pattern (Fig. 7.15c; see also Fig. 5.7) The structure of bacterial flagella led Astbury to suggest that it could be regarded as a "monomolecular muscle."

At the base of every eukaryotic flagellum and cilium is a distinct microtubular structure called the *basal body*. The architecture of the basal body is identical to that of the centriole, a structure found at opposite poles of

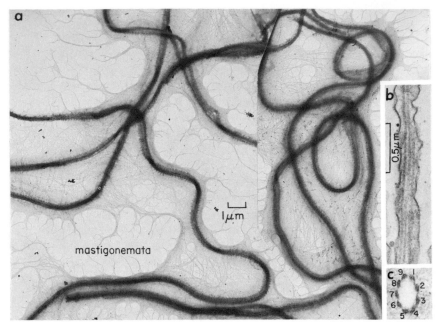

Fig. 7.14 Flagella (a) isolated from *Euglena,* not sectioned, but stained with osmium tetroxide (OsO₄). (b) Section through flagellum, longitudinal, and (c) cross-section.

the eukaryotic cell nucleus. Centrioles are found in nearly all animal cells and in the cells of many eukaryotic algae, in most higher plants, but not in certain fungi. Centrioles can come into particular prominence during mitosis.

The structural array of the basal body and the centriole is "9 + 0"; the central pair of microtubules is absent. In cells that possess mitotic centrioles, the centrioles left over from earlier cell divisions often grow projections that become flagella or cilia as the new cell differentiates. Thus, not only are basal bodies and centrioles identical in structural pattern, but centrioles can also become basal bodies. Moreover, the mitotic spindle, the structure that lies between the centrioles during cell division, is an array of microtubules composed of the protein tubulin. It is of interest from an evolutionary as well as a structural standpoint to find microtubules to be so widely distributed in all types of cells in addition to the fact that they should have a similar structure—a liquid crystalline structure.

Crystals

The cell cytoplasm possesses vacuoles, chromatophores (pigment bodies), granules, crystals, and other cellular inclusions. Many cells syn-

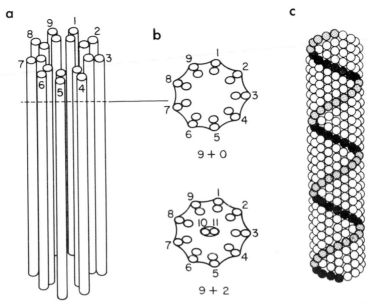

Fig. 7.15 Schematized model of (a) flagellum (holds also for cilia and centriole), (b) cross-section showing "9 + 0" and "9 + 2" microtubules, and (c) the protein (flagellin) wound as an α-helix structure.

thesize crystals, and these crystals or paracrystals can be seen in a variety of plant and animal cells (Fig. 6.2). They are also found in sex glands, the heart, and in other tissues of aged animals. The crystals that have been isolated and analyzed are primarily protein. Pigments are sometimes associated with these crystals. The crystals identified in bacteria, fungi, plants, protozoa, and animal cells have recently become of considerable interest, for it is not yet understood what function, if any, these biocrystals perform.

Receptors

As cells evolved, a variety of receptors from photoreceptors for phototropism, phototaxis, photosynthesis, and visual excitation as well as other receptors that respond to environmental stimuli developed.

The origins of these receptors may well lie with the cilia or flagella, or even the cell membrane itself, for they are differentiated structures of the cell.

The structure of various receptors, particularly the chloroplast and the visual photoreceptors as well as the effectors, muscle and nerve, will be elaborated upon in Chapters 8, 9, and 10. We hope to show relationships in their structure to function and how such relationships have analogies in liquid crystalline systems.

III. REMARKS

We know too little of the solid state and of the liquid state of matter, but we know even less of these states in the living cell. The structural studies of cells indicate that they are very complex systems of macromolecules that organized into various bodies or "organelles" that perform specific functions for the cell. These organelles consist of membranes whose molecular organization, like that of the cell membrane, is best represented by a bilayer of lipid and associated protein (Fig. 7.7).

Although considerable biochemical and structural advances have been made toward our understanding of the cell, we have not yet achieved a complete molecular description of the cell. But, as we learn more about plant and animal cells at the molecular level, we can begin to see a basic structural relationship for all living cells.

If we can draw any analogy here of our description of the living cell to liquid crystals, we can say that a necessary requirement of a living cell is that it have mobility and structural order. This is, in fact, a basic property of liquid crystals, for they have the mobility of a liquid and the structural order of a solid.

REFERENCES

Bernard, C. (1866). "Leçons sur les propriétés des tissus vivants." Bailliere et fils, Paris.
Danielli, J. F., and Davson, H. (1935). A contribution to the theory of permeability of thin films. *J. Cell. Comp. Physiol.* **5**, 495.
Green, D. E., and Young, J. A. (1971). Energy transduction in membrane systems. *Am. Sci.* **59**, 92.
Korn, E. D. (1966). Structure of biological membranes. *Science* **153**, 1491.
Lehninger, A. L. (1965). "Bioenergetics," p. 222. Benjamin, New York.
Needham, J. (1950). "Biochemistry and Morphogenesis." Cambridge Univ. Press, London and New York.
Stewart, G. T. (1974). *In* "Liquid Crystals and Plastic Crystals" (G. W. Gray and P. A. Winsor, eds.), Vol. 1, p. 308ff. Ellis Horwood Ltd., Chichestor, London.
Virchow, R. (1858). "Die Cellularpathologie in Threr Begrüdung auf Physiologische und Pathologische Gewebelehre." Hirschwald, Berlin.
Wolken, J. J. (1966). "Vision: Biochemistry and Biophysics of the Retinal Photoreceptors," pp. 59, 83. Thomas, Springfield, Illinois.
Wolken, J. J. (1967). *"Euglena,"* 2nd ed. Appleton, New York.
Wolken, J. J. (1971). "Invertebrate Photoreceptors: A Comparative Analysis," p. 22. Academic Press, New York.
Wolken, J. J. (1975). "Photoprocesses, Photoreceptors, and Evolution." Academic Press, New York.

Receptors, Effectors, and Membranes

Chapter 8
Photoreceptor Structures: The Chloroplasts

An important characteristic of life is the ability to respond to environmental changes—the stimuli, for example, responses to gravitational fields, light, temperature, touch, sound, and to chemical changes in the environment. Cells developed a variety of specialized receptors from the photoreceptors for phototropism, phototaxis, photosynthesis, and visual excitation as well as other receptors that respond to these environmental stimuli. These receptors can receive, store, transfer, and convert energy, that is, light energy to chemical, mechanical, and electrical energy—they are transducers.

The information we now have on the cell and the properties of liquid crystals permits us to consider these specialized receptors and how they are structured for function. We have already indicated that receptors are organelles of cells and that all organelles are the result of membranous processes of the cell. A central problem in these receptors, then, is to decipher their structure, to identify their receptor molecules and their molecular orientation within the receptor membranes.

Photoprocesses are so essential for life on earth that solar energy, once utilized by a living cell (for example, in photosynthesis), was an important force in evolution. Therefore it is of interest to examine the structure of the chloroplast, the photoreceptor for photosynthesis, before we discuss the visual photoreceptors and other receptor structures.

I. THE CHLOROPLAST STRUCTURE AND PHOTOSYNTHESIS

Chlorophyll is the photoreceptor molecule for photosynthesis (Fig. 5.12). Chlorophyll resides as a pigment–protein complex in the cell mem-

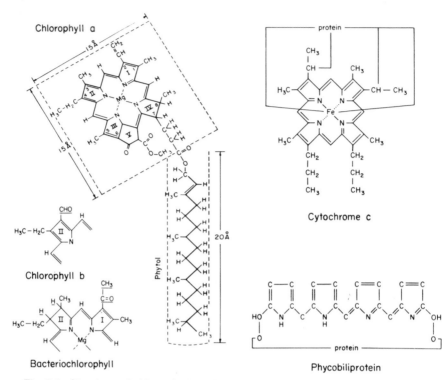

Fig. 8.1 Structure of chlorophyll *a*, chlorophyll *b*, bacteriochlorophyll, cytochrome *c*, and phycobiliprotein.

brane of the blue-green algae and in the membranes of the chloroplast, the photoreceptor structure for photosynthesis.

The structure of chlorophyll is a cyclic tetrapyrrole molecule, which has the empirical formula $C_{55}H_{12}O_5N_4Mg$. The magnesium ion is at the center of the molecule (Figs. 5.12 and 8.1). Chlorophyll exists in many isomeric forms. Of the various chlorophyll isomers, chlorophyll *a* and chlorophyll *b* are found in all higher plants. Chlorophyll *a* differs from chlorophyll *b* by possessing a methyl (-CH₃) group at the third carbon, whereas in chlorophyll *b* a formyl (-CHO) group occupies this position; chlorophyll *b* is therefore an aldehyde of chlorophyll *a* (Fig. 8.1). Chlorophyll *a* and *b* differ in absorption spectra as well as solubility. Chlorophyll *a* is present in all green plants, ferns, mosses, green algae, and euglenoids. The other chlorophyll isomers, chlorophylls *c, d,* and *e*, are found in diatoms, brown algae, dinoflagellates, crytomonads, and crysomonads. Chlorophyll *c*, which lacks a phytol group, is soluble in aqueous ethyl alcohol. Chlorophyll *d* is believed to be an oxidation product of chlorophyll *a* in which the vinyl group at position 2 is oxidized to a formyl group. It is found, together with chlorophyll *a*, in most red algae. Chlorophyll *e*, together with chlorophyll *a*, is present in small amounts in yellow-green algae (Wolken, 1975).

Bacteriochlorophyll is found free of other chlorophylls in photosynthetic purple bacteria. It differs from chlorophyll *a* in that the vinyl group at position 2 is replaced by an acetyl group, and it contains two extra hydrogen atoms at positions 3 and 4 (Fig. 8.1).

Seeds and etiolated plants (seedlings sprouted in darkness) contain no chlorophyll. However, upon exposure to light, they will turn green. The substance responsible for this reaction is *protochlorophyll,* the chlorophyll precursor (Fig. 5.12). Protochlorophyll differs from chlorophyll in that it lacks two hydrogen atoms at positions 7 and 8 in the porphyrin part of the molecule. Thus, it is an oxidation product of chlorophyll *a*.

Chemical analyses of chloroplasts indicate that they are protein, from 35 to 55%; from 18 to 37% lipids (mostly mono- and digalactosyl diglycerides and phospholipids); pigments, about 6% chlorophylls and 2% carotenoids, and from 5 to 8% inorganic matter on a dry weight basis.

The major phospholipid of photosynthetic *Euglena gracilis* is phosphatidylethanolamine. The percentage of total lipids increases from 9.1% for dark-grown *Euglena* to 13.7% for light-grown photosynthesizing *Euglena* cells (Table 8.1), which suggests the question, Is this change in lipids related to the synthesis of chlorophyll, to the development of the chloroplast, and to photosynthesis? It is interesting to note that the heat-treated

TABLE 8.1

Analysis of *Euglena gracilis* Grown in Various Environments[a]

	Photosynthesis	Chemosynthesis	
	Light-grown (%)	Dark-grown (%)	Mutant (HB) Light-grown (37°C) (%)
Product			
Protein	69.3	36.3	48.1
Lipids	13.7	9.1	7.8
α-Linolenic acid[b]	7.8	0.6	—
Carbohydrates	7.0	16.0	36.6
Ash	10.0	8.0	7.5
	100	100	100
Water	77.1	71.4	77.7

[a] Taken in part from Wolken (1967, p. 102).

[b] Data from Erwin and Bloch (1963).

(HB) mutant, which is incapable of resuming photosynthesis when placed in light, contains only 7.8% total lipid. Also, when the lipids are extracted from chloroplasts their photochemistry is inhibited, suggesting that the lipids may be more than a structural part of the chloroplasts.

In *Euglena* the galactolipids and the unsaturated fatty acids disappear when the organism is grown in darkness, and they reappear when these dark-grown cells are adapted to light again. Erwin and Bloch (1962, 1963) investigated α-linolenate content of dark-grown to light-grown *Euglena*. They found that the α-linolenic acid of dark-grown cells increased from 0.6 to 7.8% when the cells were light-grown. Upon analysis, 85% of the α-linolenic acid was found in the chloroplast fraction. Therefore, there seems to be a correlation between the photosynthetic activity and the concentration of α-linolenic acid. One of the conclusions reached was that organisms which derive energy from photosynthesis and photophosphorylation seem to require α-linolenic acid, while γ-linolenic acid, arachidonic acid, and C_{20} polynoic acids appear to be necessary for organisms that depend on respiration and oxidative phosphorylation.

The chloroplast of *Euglena* cannot develop after the organisms are grown near 37°C for more than one generation. The result is a mutation and the cell can no longer carry on photosynthesis. It does not seem unreasonable that the temperature sensitivity of the synthesis of polyunsaturated acids interferes with the development of the chloroplast, for it has been suggested that temperature variation alone, with the physiological range, can alter the physical state of fatty acids and phospholipids in the chloroplast (Table 8.1). It is of interest that phospholipids, *in vitro,* exhibit phase changes, that is, they may exist in both the lamellar form and a condensed hexagonal form, and that the transition from the lamellar to the hexagonal structure occurs near 37°C.

In chloroplasts of green plants and most algae, there is a high content of unsaturated lipids. This suggests that the transition of the physical phase of lipids in these plants would occur far below room temperatures.

II. THE CHLOROPLAST MOLECULAR STRUCTURE

Chloroplasts are the photoreceptor organelles for photosynthesis. They are referred to as *chromatophores* in photosynthetic bacteria, as *plastids* in algae (Fig. 8.2), and as *chloroplasts* in higher plants (Fig. 8.3). The chloroplasts of algae and plants are of various shapes, but generally they are ellipsoid bodies from 1 to 5 μm in diameter and from 1 to 10 μm in length. The general term chloroplasts is applied to all of these photosynthetic organelles.

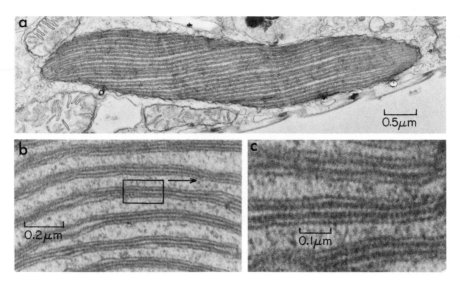

Fig. 8.2 Electron micrograph of chloroplast. (a) Longitudinal section through *Euglena* chloroplast; (b) enlarged area of chloroplast lamellae; (c) greatly enlarged lamellae. From Wolken (1975), p. 80.

Chloroplasts observed with the polarizing microscope show both form and intrinsic birefringence. Chlorophylls in the chloroplast exhibit liquid-crystalline properties (Ke and Vernon, 1971). They also exhibit measurable fluorescence. Chloroplasts posses considerable fine structure at the molecular level. Electron microscopy of chloroplasts in a variety of plants reveals that they consist of lamellae, membranes, as seen in a section through the *Euglena* chloroplast (Fig. 8.2) and the higher plant chloroplast (Fig. 8.3). The development of chloroplasts of flowering plants appears to arise from a liquid crystalline structure—that is, from a hexagonally packed form to the lamellar form seen as grana in Fig. 8.4. The elongated chloroplast of *Euglena* is seen to be made of regularly spaced lamellae. The electron dense lamellae are believed to consist of lipids, lipoproteins, and proteins because of their affinity for the osmium tetroxide used in fixation for electron microscopy. The less dense spaces are thought to contain water, enzymes, and dissolved salts. The width of this layer is more variable and appears to be dependent on its composition and temperature, whereas the electron dense fixed layers are more constant (Table 8.2).

The number of chlorophyll molecules per chloroplast, from photosynthetic bacteria to higher plants, is of the order of 10^9 molecules. The number of chlorophyll molecules is directly related to the number of lamellae (membranes) and suggests a mode of growth regulation on the mo-

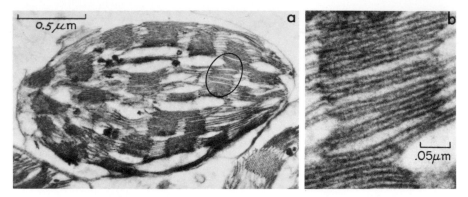

Fig. 8.3 Chloroplast of a green plant *Elodea densa*. (a) Note grana and lamellae, and (b) enlarged granum.

lecular level for development. The chlorophyll molecules would be spread as monolayers on the surfaces of lamellae as depicted in our molecular model (Fig. 8.5). This maximizes the surface area of the chlorophyll molecules for light absorption and for energy transfer to occur at specific sites in the membrane. Also, such a highly ordered structure of the chloroplast membranes provides not only for the chlorophyll and carotenoid molecules for energetic interaction, but also reactive sites for the necessary enzymatic reactions.

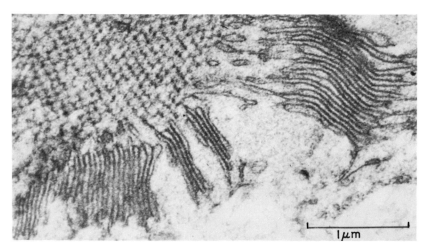

Fig. 8.4 Crystalline structure which gives rise to the lamellae in the formation of chloroplast grana as seen in green plants.

TABLE 8.2

Chloroplast Structural Data for *Euglena gracilis*[a]

Structure	Mean	Extreme
Diameter	1.23 μm	(1.04–1.42)
Length, D	6.50 μm	(5.2–9.3)
Number of dense layers, n	21	(18–24)
Dense layer thickness, T^b	242 Å	(180–303)
Interspace layer thickness, n	374 Å	(300–476)
Chlorophyll molecules, P	1.02×10^9	(0.88–1.36)
	1.34×10^9	(Calculated from absorption obtained from a single chloroplast using the microspectrophotometer)
Number of chlorophyll molecules per lamellar surface	2.5×10^7	
Number of chlorophyll molecules per cm²	4×10^{13}	

[a] Taken in part from Wolken and Schwertz (1953) and Wolken (1975).
[b] The dense layer T is a double layer: each membrane is of the order of 50–75 Å.

To establish that the chlorophyll molecules form a monomolecular layer on the surface of the lamellae was determined by calculating the area available for the chlorophyllin (porphyrin) part of the chlorophyll molecule on the lamellar surfaces. In order to do this, the geometry of individual chloroplasts (their length, diameter, and number of lamellae and their dimensions) was determined from numerous electron micrographs (Table 8.2).

The calculated cross-sectional area of the chlorophyllin part of the chloroplast molecule was found to be 225 Å² for the chloroplasts of *Euglena;* for chloroplasts in a variety of plants, the cross-sectional area in the chloroplast lamellae was found to be around 200 Å². These calculations are of the right order of magnitude for the cross-sectional area of a porphyrin molecule when spread on a water–air interface.

In the model, the chloroplast lamellae are lipid–protein membranes in which the chloroplast molecules are oriented as a monolayer on the protein surfaces whereas the phytol "tail" reaches into the lipid layer. In our model, the chloroplast lamellar network shows that four chlorophyll molecules are united to form tetrads and are oriented so that only one of the phytol tails is located at each intersection of the rectangular network (Fig. 8.5). This arrangement has the advantage of leaving adequate space for at least one carotenoid molecule for every three chlorophyll molecules. Since the molecular weight of the carotenoid molecules is one-half to two-

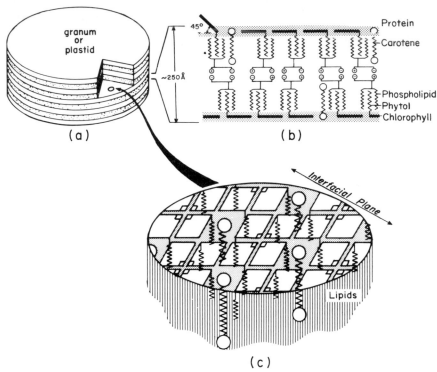

Fig. 8.5 Chloroplast molecular model (see Fig. 8.6 for the orientation of phytol in the membranes). From Wolken (1975).

thirds of the molecular weight of the chlorophyll molecules, a weight ratio of chlorophyll to carotenoid of approximately 4:1 to 6:1 would be expected. On the other hand, the carotenoid molecules are slender linear molecules, about 5 Å in diameter, and therefore more than one molecule could conveniently fit into the 15 × 15 Å holes formed by the chlorophyll tetrads. From symmetry one might expect as many as four molecules per hole, but this would lead to very tight fitting which is energetically improbable. One can therefore put a lower limit on the number of chlorophyll to carotenoid molecules of roughly 1:1 and a weight of 2:1.

It will be noted in Table 8.3 that the mono- and digalactosyl diglycerides account for the major lipids on the chloroplast, and the lamellae of the chloroplasts may be associated with the structure of the lipid molecules. Benson (1966) suggests that these lipids, because of their properties, can form a lipid or lipoprotein matrix for the chlorophyll monolayers. From spatial considerations the ratio of two galactosyl diglyceride molecules to

TABLE 8.3

Composition of Spinach Chloroplast Quantasome[a]

Number of molecules		
230	Chlorophylls	
	160	chlorophyll *a*
	70	chlorophyll *b*
48	Carotenoids	
9380	Nitrogen atoms as proteins (approximately 426 protein molecules)	
624	Lipids	
460	⌠114	digalactosyl diglyceride
	⌡346	monogalactosyl diglyceride
	116	phospholipids
	48	sulfolipids

[a] Taken in part from Parks and Biggins (1964).

one chlorophyll molecule could stabilize all the chlorophyll molecules in the monolayer. That is, there would be one phytol chain of chlorophyll for four cis-unsaturated acyl chains of galactosyl diglyceride (Rosenberg, 1967). Such a relationship is illustrated in Fig. 8.6 and fits in with the proposed molecular model for the chlorophyll in the chloroplast lamellae (Fig. 8.5).

Other molecular models for the chloroplast have been proposed, for there are several possible ways in which the chlorophyll molecules could be oriented in the chloroplast lamellae. If the porphyrin parts of the chlorophyll molecules lie at 0° as depicted in Fig. 8.5a, their greatest cross-section would be available. If they are oriented at increasing angles to 45° (Fig. 8.5b), the cross-sectional area would be decreased to about 100 Å². Since the chlorophyll molecules in the chloroplast are in a dynamic state, they would arrange themselves for maximum light absorption, so their greatest cross-section is available for light capture.

III. CHLOROPLASTIN: CHLOROPHYLL–PROTEIN MICELLES

We have already noted that chloroplasts contain large quantities of lipids (on the order of 35%), and it is difficult to solubilize them in aqueous media, but molecular dispersions can be obtained by extraction with low concentrations of detergents. Of the non-ionic detergents, digitonin, ($C_{55}H_{90}O_{20}$), a digitalis glycoside (Fig. 8.7), resembles in structure choles-

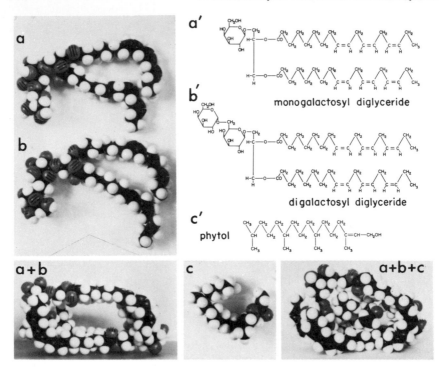

Fig. 8.6 Possible molecular association between the mono- and digalactosyl diglycerides (the four cis-unsaturated acyl chains) and the phytol of the chlorophyll molecules as illustrated in (a), (b), (c). (Adapted from Rosenberg, 1967.) From Wolken (1975, p. 84).

terol (Fig. 5.4). The structure of the digitonin micelle system is liquid crystalline since it is ordered in only two dimensions.

Digitonin has a strong attraction for dye molecules and has been most useful in extracting the visual complex rhodopsin from the retinal rods of eyes, from which must of our knowledge of the photochemistry of vision has been obtained. Chloroplasts can also be extracted with 1.8% digitonin. When such an extract is further separated by high-speed centrifugation, a fraction is obtained that is a clear green solution. This is known as chloroplastin. It is believed that in the process of extraction, digitonin opens the pigment–lipid–protein of the chloroplast lamellae to form chloroplastin micelles (Fig. 8.8).

The chloroplastin absorption spectrum is similar to that of the chloroplast. The relative ratio of chlorophyll, lipid, and protein is similar to that found in the analysis of the chloroplast. Chloroplastin also exhibits bire-

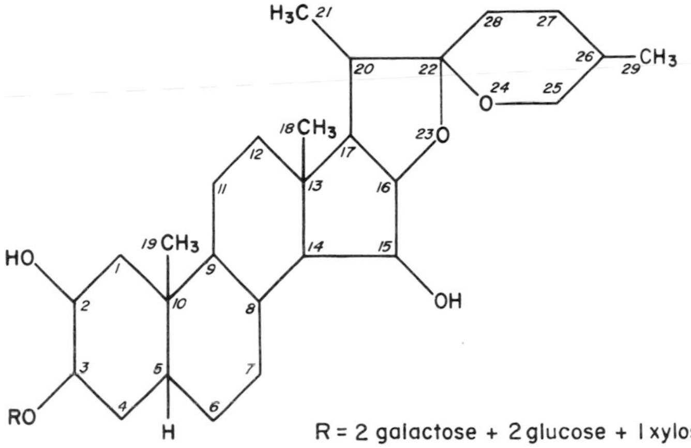

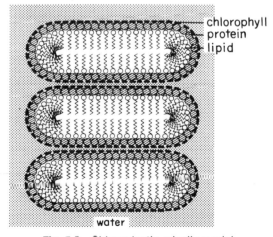

Fig. 8.7 Structure of digitonin. Compare to cholesterol, from Fig. 5.4.

R = 2 galactose + 2 glucose + 1 xylose

fringence, indicating that there is an ordered alignment of the molecules analogous to that of a liquid crystal.

Analytical data show that there is one chlorophyll molecule associated with one protein molecule in the chloroplastin with an estimated molecular weight of 20,000 to 40,000 (Wolken, 1967). Unfixed and unstained chloroplastin viewed in the electron microscope shows particles which range in diameter from 100 to 1000 Å. Assuming that each chloroplastin liquid crystalline structure is about 200 Å in diameter, it could contain

Fig. 8.8 Chloroplastin micelle model.

about 225 chlorophyll molecules, 55 carotenoid molecules, one molecule of cytochrome, and one ferredoxin molecule together with protein and lipid, the estimate for a functional photosynthetic unit.

Chloroplastin, the chlorophyll complex in aqueous 1–2% digitonin, when caused to flow through a capillary, becomes birefringent when observed through crossed polarizers; hence there is an alignment of the molecules in solution. If a drop of chloroplastin is evaporated rapidly from a surface, periodic structures will also be formed similar to those just described (Fig. 8.9). When these chloroplastin structures are scanned with the microspectrophotometer at the major absorption peak for chlorophyll (675 nm), chlorophyll is found within the rings (Fig. 8.10) and not in the interspaces (compare to electron micrographs of fixed chloroplasts, Figs. 8.2 and 8.3). These experiments indicate that in chloroplastin, chlorophyll molecules become oriented in a liquid crystalline-like structure.

Therefore, the chloroplastin structure provides a good model for studying photosynthesis outside the living cell. It was found that such systems can photoreduce a dye, evolve oxygen, and in the presence of the right cofactors, perform some of the primary steps of photosynthesis, turning

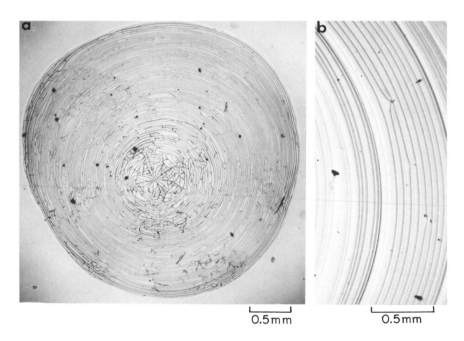

Fig. 8.9 Digitonin, 1.8% saturated solution, showing periodic wave formation from a drop of solution upon rapidly removing the water (a). (b) Enlarged area to show the periodic rings.

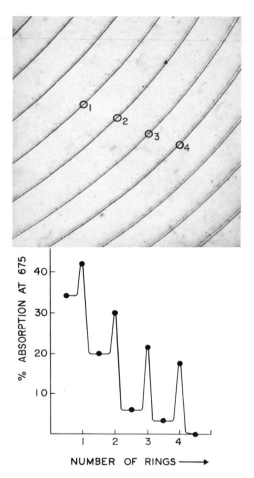

Fig. 8.10 Chloroplastin extracted with digitonin (as in Fig. 8.8) showing that chlorophyll is located in the digitonin rings. From Wolken (1967), *"Euglena,"* p. 139.

inorganic phosphate to organic phosphate (ATP) outside the living cell (Wolken, 1967, 1975).

IV. REMARKS

If any analogy is to be drawn between the chloroplast structure and liquid crystals, we must look closely at the chemistry and physics of their membranes. Since photosynthetic reactions take place in the lipoprotein matrix of the chloroplast membranes, the physical state of their membrane lipids plays an important role in the functioning of the photosynthetic reactions. Therefore the structural relationship of the chlorophyll–lipid–protein molecules that comprise the chloroplast membranes suggest a liquid crystalline system (Wolken, 1967; Ke and Vernon, 1971).

The finding that chlorophyll molecules orient in liquid crystals (Journeaux and Viovy, 1978) is very suggestive as a model system for further investigation concerning the photoactivity of chlorophyll in organized structures, and hence in photosynthesis.

REFERENCES

Benson, A. A. (1966). On the orientation of lipids in chloroplast and cell membranes. *J. Am. Oil Chem. Soc.* **43**, 265.

Erwin, J., and Bloch, K. (1962). The linolenic acid content of some photosynthetic microorganisms. *Biochem. Biophys. Res. Commun.* **9**, 103.

Erwin, J., and Bloch, K. (1963). Polyunsaturated fatty acids in some photosynthetic microorganisms. *Biochem. Z.* **338**, 496.

Journeaux, R., and Viovy, R. (1978). Orientation of chlorophylls in liquid crystals. *Photochem. Photobiol.* **28**, 243.

Ke, B., and Vernon, L. (1971). Living systems in photochromism. *In* "Photochromism" (G. H. Brown, ed.), p. 687. Wiley (Interscience), New York.

Parks, R. B., and Biggins, J. (1964). *Science* **144**, 1009.

Rosenberg, A. (1967). Galactosyl diglycerides: Their possible function in *Euglena* chloroplasts. *Science* **157**, 1191.

Wolken, J. J. (1967). *"Euglena;"* 2nd ed. Appleton, New York.

Wolken, J. J. (1975). "Photoprocesses, Photoreceptors and Evolution." Academic Press, New York.

Wolken, J. J., and Schwertz, F. A. (1953). Chlorophyll monolayers in chloroplasts. *J. Gen. Physiol.* **37**, 111.

Chapter 9
Visual Systems: The Optical and Photoreceptor Structures

I. THE EYE AND VISUAL EXCITATION

How the eye functions in vision is of major importance to man. The gross anatomy of the human eye is shown schematically in Fig. 9.1. The eyeball, which is approximately spherical, houses the complete optical apparatus. Image formation takes place by the refraction of light on one or more spherical surfaces that separate media of different refractive indexes. These two refracting structures are the cornea (refractive index 1.336) and the lens (refractive index 1.437). A variable aperture, a diaphragm, is provided by a contractile membranous partition, the iris, that regulates the size of the light opening, the pupil. In dim light, in order to admit more light through the lens, the pupil opens to an extent governed by the activity of the retina. The lens also acts as a filter by sharply cutting off the far edge of the ultraviolet region at about 360 nm.

A. The Retina

The vertebrate retina is a complex structure which is closely attached to the pigment epithelium. The nervous cell layers comprise the rod and cone cells, the bipolar cells, and the ganglion cells. The photoreceptors, the rods and cones, are arranged in a single-layered mosaic array (Fig. 9.2). It is estimated that in the human retina there are about 1×10^8 retinal rods and 7×10^6 retinal cones. The rods and cones are connected with a highly developed system of connecting and interconnecting neurons.

B. The Retinal Photoreceptors

The rods and cones are specialized photoreceptor cells of the retina. Each photoreceptor has an inner segment and a rod- or cone-shaped outer segment which contains all the photosensitive visual pigment, rhodopsin.

119

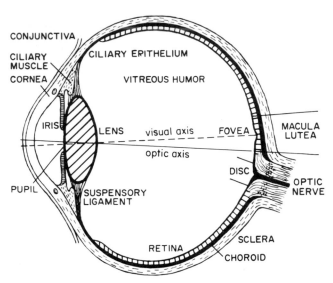

Fig. 9.1 Structure of human eye.

The inner segments of the rods and cones are densely packed with mito-chondria. Communication between the inner segment and the outer seg-ment occurs by a cilium or flagellum (Figs. 7.14 and 7.15) which passes from the inner segment into the outer segment (Figs. 9.2 and 9.3).

We have illustrated the human and bovine retinal rods in Figs. 9.2 and 9.3. What is apparent in these retinal rod outer segments is that they are comprised of lamellae. As an example of the vertebrate rod, let us ex-amine in more detail the retinal rod structure of the frog, an amphibian. The retinal rod outer segments of the frog can be severed from the retina simply by shaking and are easily observed with the light microscope. The outer segment is about 6 μm in diameter and 60 μm in length (Fig. 9.4a). The rod outer segments, when observed with the polarizing microscope, are birefringent. Schmidt (1935) noted that when the rods were extracted with organic solvents to remove the lipids, there was a reversal in sign of the birefringence. To account for these optical changes, Schmidt (1937, 1938) suggested that the lipid molecules lie parallel to the axis of the outer segment and that there is nonlipid material arranged at right angles to the long axis. When the rods were treated with dilute acids or alkali, the bire-fringence disappeared, accompanied by a lengthening of the rod to as much as ten times the original dimension. Further observations on the dichroism of the outer segment suggested that the visual pigment was present in the nonlipid regions of the rod. With the rods free from lipid, Schmidt was able to show a complete curve for birefringence as a function

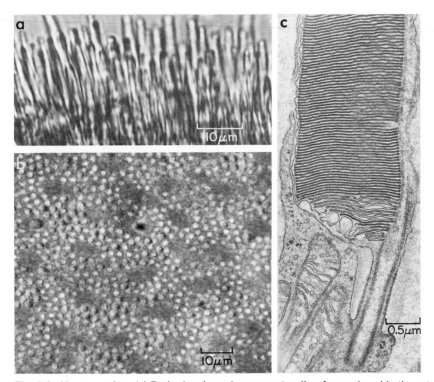

Fig. 9.2 Human retina. (a) Retinal rods and cones extending from a bend in the retina. (b) Surface view of retinal rods and cones. Note array of rods (rhabdomlike arrangement). From Wolken (1966), p. 23. Reprinted by permission. (c) Human retinal rod, showing the outer segment lamellae and inner segment of the retinal cell. (Courtesy of Dr. T. Kuwabara.)

of the refractive index of the medium. As a result, he postulated that the rod outer segments are highly ordered structures of lipids and proteins. This is borne out by the data in Table 9.1 and by electron microscopy, for all vertebrate retinal rods and cones that have been studied by electron microscopy show that they are structured of lamellae, double membraned disks of the order of 250 Å in thickness, as illustrated for human, bovine, and frog retinal rods (Figs. 9.2, 9.3, and 9.4).

C. Invertebrate Photoreceptors

The invertebrates are the most numerous as well as diverse of all animal species, and their visual organs are equally diverse. In examing the photoreceptor structures from the protozoa to coelenterates, flatworms, roundworms, and segmented worms, we find structures ranging from eye-

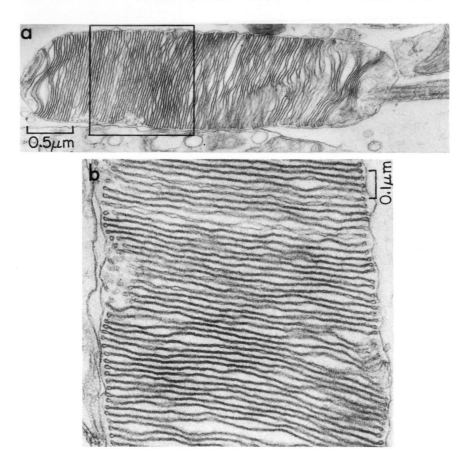

Fig. 9.3 Bovine retinal rod. (a) Longituainal section. (b) Enlarged area in the retinal rod to show double-membraned lamellar discs. From Wolken (1971, p. 98).

spots to simple pinhole type eyes. For example, in the protozoan flagellate *Euglena* (Figs. 7.4 and 7.6), a photoreceptor system of eyespot, para-flagellar body, and flagellum evolved for phototaxis (Fig. 9.5). The para-flagellar body is the photoreceptor structure. The organism can be likened to a photoneurosensory cell, for it possesses the structures and chemistry for such a cell (Wolken, 1971, 1975, 1977). Electron microscopy of the para-flagellar body and optical diffraction of the electron micrographs indicate that it is a lamellar structure of packed rods. The twist in the lamellae re-sembles the helicoid structure described for the cuticle (Fig. 7.6) and can be considered in its formation a cholesteric–nematic liquid crystal.

In the earthworm, an annelid, there are light-sensitive cells that contain a lens embedded in the body wall. In many leeches, such photosensory

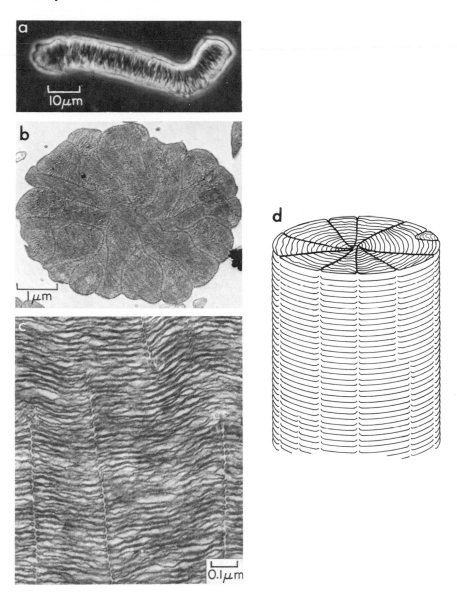

Fig. 9.4 Frog retinal rod. (a) Isolated retinal rod outer segment. (b) Cross section of outer segment. (c) Longitudinal section. (d) Schematic drawing of outer segment of retinal rod. Note structural relationship of fused rhabdom of arthropods (Figs. 9.9–9.11). From Wolken (1971, p. 97; 1975, p. 187).

TABLE 9.1

Comparative Composition of Proteins and Lipids
in Retinal Outer Segments[a]

	Dry weight (%)	
	Cattle	Frog
Total lipid	38.15	40.6
Total protein	61.85	59.4
	Total lipids (%)	
Phosphatidylethanolamine	38.5	25.2
Phosphatidylserine	9.2	9.5
Phosphatidylcholine	44.5	49.4
Sphingomyelin	1.3	1.8
Other phospholipids	6.5	9.2

[a] Taken from published data from various sources and averaged. Refer also to Table 7.1.

cells are gathered into cups shielded by pigment and covered by a specialized area of the animal's cuticle which functions as the lens.

D. The Compound Eye

The arthropods, which include arachnids, crustacea, and insects, all have image-forming compound eyes whose history goes back more than 500 million years to the trilobites, as seen in the fossil *Phacops* (Fig. 9.6). The image-forming compound eye developed by arthropods is very efficient for detecting movement in their visual field and for short-range vision. Arthropods can orient to the direction of vibration of polarized light, which indicates a polarized light analyzer within the eye. Many arthropods discriminate colors and therefore have visual pigments with different absorption maxima. Let us examine the compound eyes of a few selected insects and crustacea to see how they are structured.

The compound eye is composed of numerous ommatidia, or eye facets, ranging from a few in the ant to hundreds in the dragon fly. Each ommatidium is a complete eye, consisting of an optical system, a corneal lens, a crystalline cone, and a photoreceptor system of retinula cells (Fig. 9.7). The photoreceptor is the rhabdomere, a differentiated structure of the retinula cells. Collectively, the rhabdomeres form the rhabdom, or the retina.

The anatomically distinct types of insect compound eyes are the "apposition" eye and the "superposition" eye. The apposition type of eye is

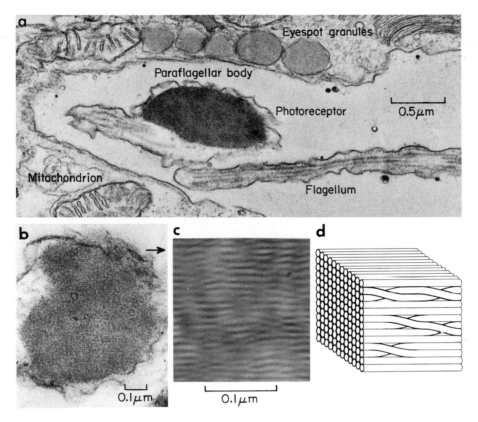

Fig. 9.5 *Euglena gracilis*. (a) Eyespot area for phototaxis, eyespot granules, paraflagellar body, and flagellum. (b) Paraflagellar body, the photoreceptor. (c) Paraflagellar body, filtered image obtained by optically diffracted electron micrograph. (d) Schematic drawing of orientation of fibers (lamellae) in the paraflagellar body. From Wolken (1971, p. 27; 1977). Reproduced by permission.

one in which the rhabdomeres forming the rhabdom lie directly beneath or against the crystalline cone (Fig. 9.7a). This type of compound eye structure was believed to be characteristic of diurnal insects. The superposition type of eye is one in which the rhabdom resides at a certain distance away from the crystalline cone (Fig. 9.7b). The superposition eye was believed to be characteristic of nocturnal insects, and the superposition mechanism was thought to be important for the requisite increase in light-gathering power. However, superposition eyes have been found in both diurnal and nocturnal species.

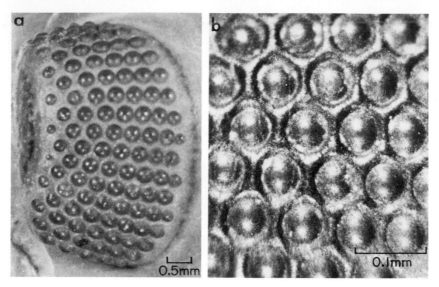

Fig. 9.6 *Phacops rana,* trilobite fossil compound eyes. (a) From Devonian Silica Shale, New York. (b) Enlarged area in compound eye. From Devonian Silica Shale, Ohio. These compound eyes corneal lens structure consist of calcite and chitin. (Fossils courtesy of Carnegie Museum of Natural History, Pittsburgh, Pennsylvania.) From Wolken (1975, pp. 140–141).

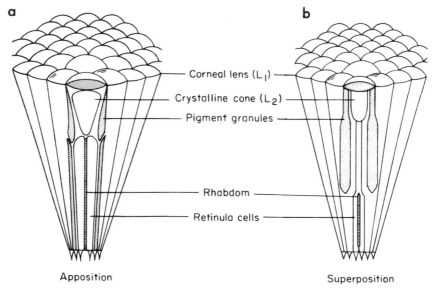

Fig. 9.7 Compound eye structure, showing eye facets, ommatidia. (a) Apposition type compound eye. (b) Superposition type compound eye.

E. The Photoreceptor System

The number of rhabdomeres that form the rhabdom varies; in the open-type rhabdom, there are usually five to seven, with an asymmetric rhabdomere lying in the same plane (Fig. 9.8). In the closed-type rhabdom, there are four to eight rhabdomeres forming a symmetrical arrangement, whereas its asymmetric rhabdomere lies in another plane (Figs. 9.9 and 9.10). The closed-type rhabdom is most likely an efficiency mechanism used for light capture by nocturnal insects. In general, nocturnal insects such as cockroaches, fireflies (Fig. 9.9), and moths have rhabdoms that occupy a large cross-sectional area as compared to daylight-active insects (Fig. 9.8; Wolken, 1971). The fine structure of all rhabdomeres is that of packed microtubules from about 300 to 500 Å in outside diameter whose double-walled membranes are from 50 to 100 Å in thickness (Figs. 9.11 and 9.12). In the rhabdom, the microtubules of two adjacent rhabdomeres are oriented perpendicular to each other, whereas in the opposite pair of rhabdomeres, the microtubules are parallel. They can appear as either tubules with hexagonal packing or lamellae, depending on the angle of the cut through the rhabdomere.

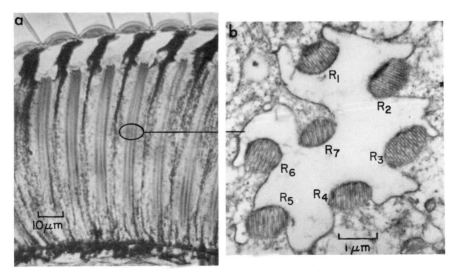

Fig. 0.0 Compound eye, *Drosophila melanogaster*. (a) Longitudinal section through several ommatidia, showing the corneal lens, crystalline cone, rhabdom, and migrating pigment granules. (b) Cross section through the rhabdom, to illustrate the orientation of the rhabdomere (photoreceptors), R_1–R_7. From Wolken (1971, p. 53).

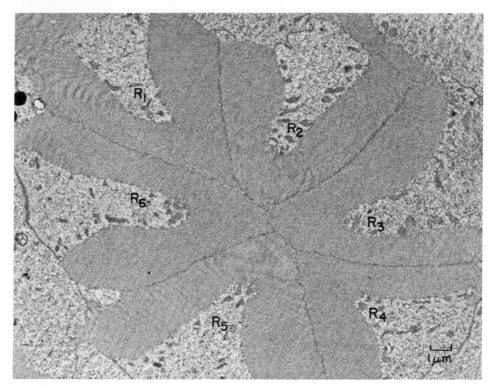

Fig. 9.9 The firefly (*Photuris pennsylvanica*) rhabdom; cross section through the rhabdom showing geometric arrangement of its rhabdomeres R_1–R_6. From Wolken (1971, p. 62).

F. Tracheoles

In certain insects, tracheoles surround the rhabdoms. They are seen in the June beetle (Figs. 9.10 and 9.13a). According to Miller and Bernard (1968), the ridges of the tracheoles function as a quarter-wave interference filter. This is possible because the tracheoles consist of the cuticle polymer chitin, which is formed in a twisted lamellar pattern, and the thickness and spacing of the lamellae correlate with the reflected colors. The tracheoles surrounding the rhabdoms of moths are responsible for the eye glow of these insects, and in butterflies, this is seen as a red glow. Tracheoles are also found in the photocytes, the luminescent cells (the lantern) in the firefly tail (Fig. 9.13b).

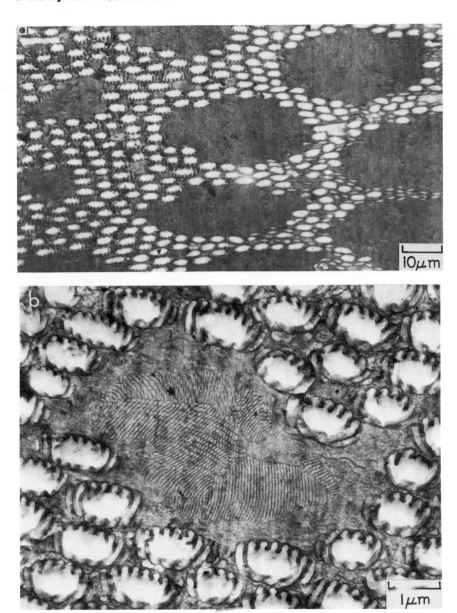

Fig. 9.10 June beetle (Scarab, *Phyllophaga*). (a) Cross section through compound eye in photoreceptor area, showing rhabdoms surrounded by tracheoles. (b) Higher magnification; cross section of a rhabdom and tracheoles. From Wolken (1971, pp. 66, 67).

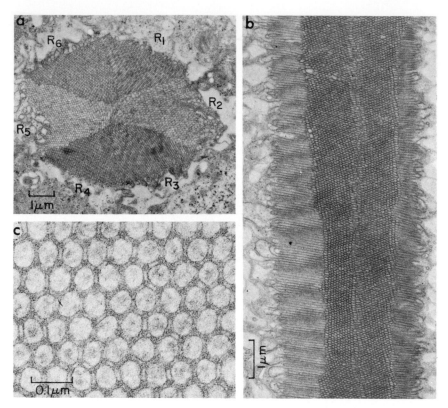

Fig. 9.11 The carpenter ant (*Camponotus herculenus pennsylvanicus*) rhabdom. (a) Cross section through the rhabdom, showing rhabdomeres R_1–R_6. (b) Longitudinal section of the rhabdom. (c) A higher resolution of a rhabdomere, showing microtubules packed as hexagons.

G. Polarized Light Analysis

One of the interesting phenomena exhibited by animals with compound eyes is that they are able to analyze the plane of vibration of polarized light. They use it as a compass for navigation; the direction of polarization indicates the relative direction of the sun. Sir John Lubbock (1882), an English banker, wondered how ants find their way to and from their nests while foraging for food. His observations led him to conclude that ants use the sun as a compass for orientation and navigation. Although this seemed to be a reasonable explanation, it was soon questioned: how could insects find their way when the sun was hidden from view? The question was not really answered until von Frisch's (1949) behavioral studies of the bee suggested that, in addition to the direction of a point

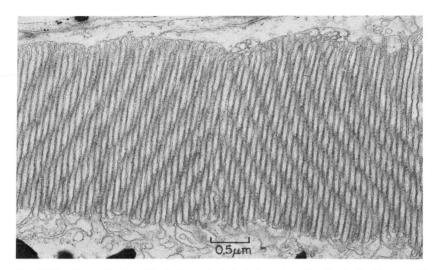

Fig. 9.12   The waterflea (*Daphnia pulex*) rhabdom showing the helical twist in the geometry of the rhabdomere.

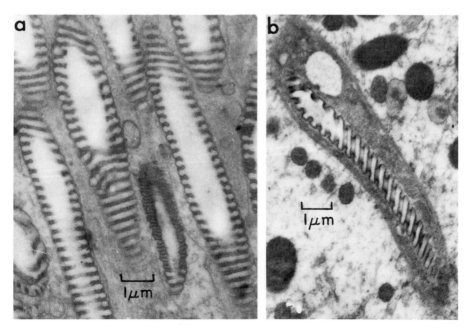

Fig. 9.13   Tracheole structure, longitudinal sections in (a) June beetle (Scarab, *Phyllophaga*) in rhabdom photoreceptor area; (b) firefly (*Photuris pennsylvanica*) in photocyte cells (bioluminescent cells) of the lantern. From Wolken (1975; p. 271).

source, the sun, insects could utilize polarization information from a patch of blue sky. To test this hypothesis, von Frisch (1950, 1967) constructed a model using eight triangular polarizing elements, each transmitting a quantity of light proportional to the degree of polarization. In this model, opposite pairs of rhabdomeres have their polarizers in parallel orientation. The arthropod and mollusc photoreceptor's fine structure as revealed by electron microscopy, and the striking geometric arrangement of perpendicular and parallel microtubules that form the rhabdom would support von Frisch's model (Figs. 9.8 and 9.9).

Another way that arthropods detect polarized light is that the analyzer mechanism could be based on the dichroism of the visual pigment, rhodopsin, within the rhabdomeres. If, in fact, the rhabdom constitutes a dichroic analyzer, its properties would depend on the rhodopsin molecules oriented with their major axes parallel to the tubule direction and hence perpendicular to the normally incident illumination. Based on this, Waterman *et al.* (1969) provided an explanation of the polarized light analyzer action in the rhabdom that indicated that the absorbing dipoles of the rhodopsin molecules, as in the vertebrate retinal rods, lie parallel to the membrane surface but are otherwise randomly oriented.

A number of other hypotheses for interpreting the mechanism of directional analysis of polarized light have been presented. One is that the direction of the electric vector of a beam of plane polarized light can be perceived through simple intensity discriminations because the direction of vibration is resolved into intensity gradients by reflection from the background. In fact, reflection patterns from the environment do resolve polarized light into patterns of graded intensity. Another suggestion involved a model that is dependent upon reflection refraction at the air–corneal lens interface. This then brings us to the optics and structure of the corneal lens.

H. The Photoreceptor Optical System

The optical system of compound eyes consists of the corneal lens (L_1) and the crystalline cone (L_2), as illustrated in Fig. 9.7. In certain insects such as the firefly and June beetle, the corneal lens extends into the region normally occupied by the crystalline cone (Fig. 9.14). Compound eyes so structured are referred to as either exocone or pseudocone type of ommatidia. In crustacea, for example, the waterfleas, the compound eye is contained with the cuticle of the animal. As a result, the cuticle serves as a common lens for all the ommatidia of their compound eye.

Since the corneal lens is derivable from the cuticle, let us briefly review its structural properties. Chemically, the cuticle is almost all chitin, a

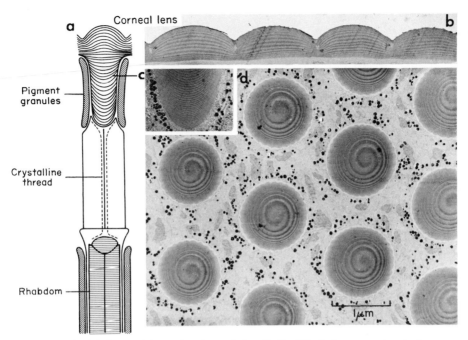

Fig. 9.14 (a) Corneal lens structure of the firefly (*Photuris pennsylvanica*) compound eye. Model of the ommatidium (a) in the compound eye. (b), (c), (d) Electron micrographs of lens structure. From Wolken (1971, pp. 59, 60).

polysaccharide polymer (Fig. 5.11). The structural and optical properties of insect cuticles were studied by Neville and Caveney (1969) and Neville (1975). From light microscopy and electron microscopy, the cuticle was found to consist of lamellae that have a characteristic fibrous pattern. Each of the lamella observed in oblique section seemed to be composed of fibrils arranged in an arced or parabolic pattern. The direction of the fibrils appears to be rotated 180° from one lamella to the next. The cuticle structure in its formation is likened to that of a cholesteric–nematic liquid crystal. Bouligand (1972, 1974) has also drawn a comparison between the cuticle fibrous structure and other biological structures and concludes that twisted fibrous biological structures have many geometrical and optical properties in common with cholesteric–nematic liquid crystals, because the orientation of the molecules in cholesteric–nematic liquid crystals is in parallel planes with the planes twisting to form a helicoid (Fig. 3.2c).

Electron microscopy of the corneal lens of the firefly (*Photuris pennsylvanica*) shows that it is formed of a laminated series of paraboloids, and the fibers of each lamella appear to be rotated as in the cuticle structure

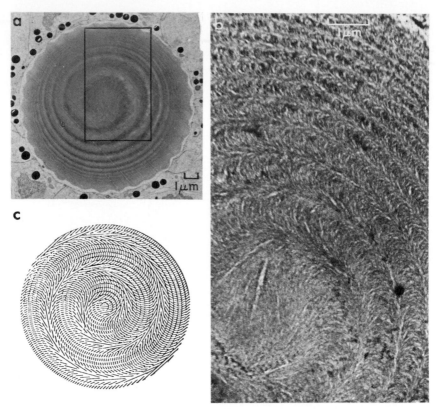

Fig. 9.15   (a) Corneal lens of firefly (*Photuris pennsylvanica*). (b) Enlarged section of (a). (c) Diagrammatic representation, projection on the cutting plane of the fibril directions in the concentric rings.

(Figs. 9.14 and 9.15). This is also seen in the corneal lens of the June beetle (Scarab, *Phyllophaga*). In transverse section, lamellae form a single or double spiral (Fig. 9.16). Similar spiral structures were then observed for the carpenter ant (*Camponatus herculenus*), the fruitfly (*Drosophila melanogaster*), and the housefly (*Musca domestica*). A structural model for the corneal lens of these insects is illustrated in Fig. 9.17.

The corneal lens has the right spacing, pitch in the visible range, to function as a polarizer of visible light, whereas the rhabdom, because of the way it is structured, would serve as the analyzer for the polarized light and could function as a navigational device for these animals.

The helical structure is probably a common architecture found from molecules to biological structures. We have previously mentioned that the polynucleotides, the polypeptide polybenzyl-L-glutamate, DNA, and

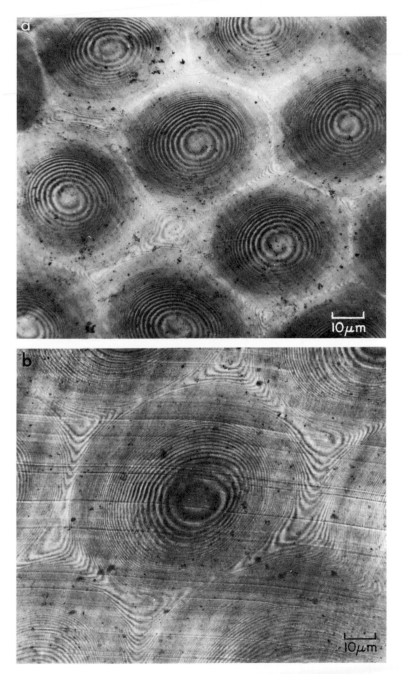

Fig. 9.16 Corneal lens structure of June beetle (Scarab, *Phyllophaga*). (a) Cross section through many lenses. (b) Higher magnification of cross section of lens.

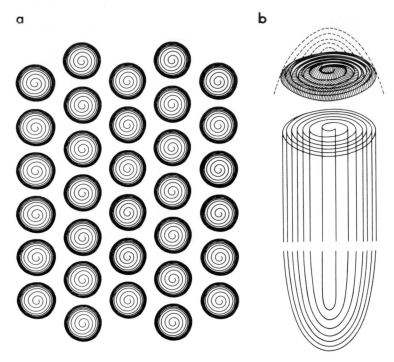

Fig. 9.17 Model for corneal lens of arthropods, showing various representations of the helix.

RNA are structured into a helix and are liquid crystalline. In addition to the cuticle and corneal lens of compound eyes described here, such a helical arrangement has been noted for ascidian tunics, plant cell walls, connective tissue of invertebrates, and vertebrate cartilage bone. It is of interest to note for comparison the helical structure of the liquid crystal (Fig. 9.18) to the helical structures of the corneal lens (Figs. 9.14–9.16).

II. THE VISUAL PIGMENT

A. Rhodopsin

We have already pointed out that the rhodopsin molecules are in or on the membranes of the lamellae of the vertebrate retinal photoreceptors and are similarly ordered in the microtubules of the invertebrate photoreceptor rhabdomeres. However, additional information on the visual pigment will be helpful in trying to understand how the photoreceptors are structured for visual excitation.

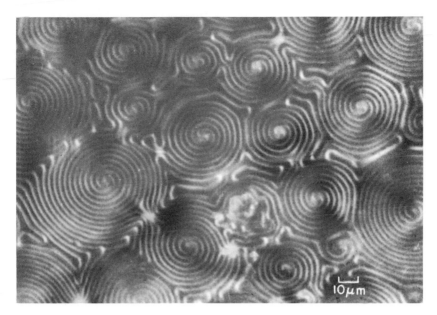

Fig. 9.18 *p*-Methoxybenzylidene-*p-n*-butylaniline + 10% (by weight) cholesteryl nonanoate. Surface disclinations 20°C, darkfield. (Courtesy of Alfred Saupe.) Cf. Fig. 6.4 and Figs. 9.14–9.16.

B. Retinal

All visual photoreceptor molecules thus far isolated from vertebrate and invertebrate eyes are rhodopsins which contain retinal$_1$ or retinal$_2$ (derivable from vitamin A_1 or vitamin A_2) as the chromophore (Fig. 9.19). However, retinal has been identified in the very halophilic bacteria, for example, *Halobacterium halobium*. In these bacteria, rhodopsin is located in its cell's "purple" membrane, much like that in the visual photoreceptor membranes. It is thought that rhodopsin in these bacterial membranes functions as a proton pump in the metabolic process of photophosphorylation, more closely related to photosynthesis than to visual excitation. The fact that retinal and its complex rhodopsin is found in organisms without eyes suggests a more generalized biological function than simply that of vision.

C. Opsin

The protein opsin, with retinal, forms the visual complex rhodopsin. Rhodopsins isolated from different species have different absorption spectral peaks and solubilities. Since opsins do not themselves absorb

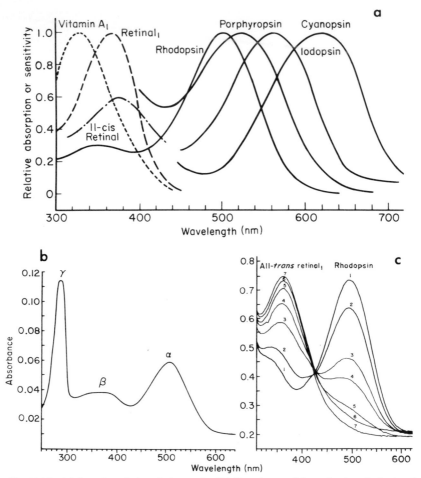

Fig. 9.19 (a) Spectra of visual pigments. (b) Spectrum of frog rhodopsin (extracted in 4% tergitol), the absorption peak around 280 nm is due to the protein, opsin. (c) Spectra upon bleaching frog rhodopsin to all-*trans*-retinal. From Wolken (1966), pp. 40–43. Reprinted by permission.

light in the spectral range 480–560 nm, the shifts in the positions of the visual pigment absorption maxima must be attributed to the effect of the opsins on the spectral properties, the absorption maxima of the various rhodopsins (Fig. 9.19).

Studies of opsin by Raman spectroscopy indicate that part of the native opsin skeletal structure is that of an α-helix. Other studies also indicate an α-helix structure, with no detectable β-structure, for both opsin and rhodopsin. Furthermore, Raman spectroscopy of opsin membranes reveals details about the conformation of the protein opsin and the membrane phospholipids; temperature studies indicate that the membrane lipids remain fluid to at least 15°C, that of a gel to a liquid crystal transition. Such low transition temperature could be consistent with a high degree of unsaturated lipids present in the opsin-containing membranes, the lamellae of the retinal rods (Rothschild *et al.*, 1976).

D. Rhodopsin in Solution

Extracting rhodopsin from photoreceptor membranes is difficult because of its insolubility in water, but rhodopsin can be extracted by surfactants. Most of the photochemistry of rhodopsin has been carried out with such rhodopsin solutions in 1–2% aqueous digitonin. One of the interesting properties of these surfactants is that they are lyotropic liquid crystalline systems. Another property is a strong affinity for dyes such as chlorophyll and retinal. This is seen when a drop of rhodopsin solution is allowed to evaporate on a microscope slide. A regular pattern of lamellae is formed. When these lamellae are scanned with the microspectrophotometer at 500 and 375 nm, the absorption maxima of rhodopsin and retinal, respectively, the pigment is found to be associated with the lamellae. This is simila. to that observed for chlorophyll in chloroplastin (Fig. 8.10). When such a film of rhodopsin is placed on a grid and is examined in the electron microscope, the lamellae are equidistant in spacing, but there are various types of substructures that appear crystalline, as seen in Fig. 9.20. Such substructures are similar to those found in liquid crystalline systems (Luzzati *et al.*, 1968).

It is conceivable then that the photoreceptor structure origin lies with the formation of a liquid crystalline state and becomes stabilized by the pigment molecules.

E. Retinal Rod Molecular Structure

The structure of all vertebrate retinal rod outer segments are double-membraned lamellae, or discs (Figs. 9.2–9.4). The membranes of the lamellae are from 50 to 75 Å thick, separated by less dense material, and the discs, or bilayers, are of the order 250 Å in thickness.

Knowing the geometry of the retinal rods, their length, diameter, the number of lamellae, and the number of rhodopsin molecules per rod, we can hypothesize a molecular model for a retinal rod. To do so, several assumptions are necessary. These are (1) that the retinal rod lamellae are membranes of protein and lipid, (2) that the rhodopsin molecules are on or in the protein of the membrane as a monomolecular layer, and (3) that there is one molecule of retinal per opsin in the rhodopsin complex. These assumptions are supported by chemical analysis (Wolken, 1966) and

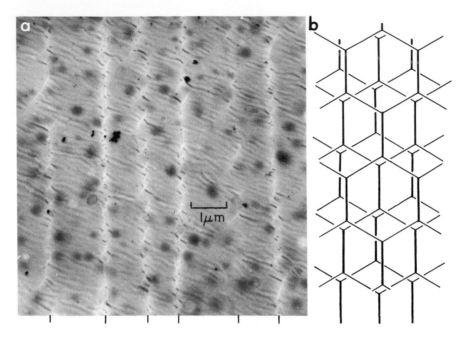

Fig. 9.20 (a) Film of bovine rhodopsin in digitonin (1.8%). Electron micrograph negatively stained with phosphotungstic acid (PTA). (b) Substructure in the lamellae.

shown in Tables 9.1 and 9.2. With these data we have calculated the surface area that the rhodopsin molecules would occupy in the lamellae, the diameter of the rhodopsin molecule, and the molecular weight for bovine and frog rhodopsin (Table 9.2). The molecular structure of the rod outer segment, based on these data and calculations, is shown in Fig. 9.21. X-Ray diffraction studies of the frog outer rod segment seem to support such a model (Blaurock and Wilkins, 1969; Blasie, 1972). A small area is enlarged to show the molecular packing of retinal with opsin on the lamellar membranes. In rhodopsin, retinal is complexed with the protein opsin through lysine in opsin in a Schiff-base linkage.

The orientations of retinal with respect to opsin in the membrane is not precisely known and could be in either orientation, as shown in Fig. 9.21. Upon light excitation of rhodopsin, there is a conformational change in opsin in the membrane which releases the 11-*cis*-retinal to the all-*trans*-retinal, and excitation occurs. In the dark, rhodopsin is restored to its original state.

TABLE 9.2

Structural Data for Retinal Rod Outer Segment[a,b]

	Frog	Cattle
Average diameter, D (μm)	5.0	1.0
Thickness of lamellae, T (Å)	200	200
Number of lamellae per rod, n	1000	800
Number of rhodopsin molecules per rod, N	3.8×10^9	4.2×10^6
Calculated cross-sectional area of rhodopsin (Å^2)	2620	2500
Calculated diameter of rhodopsin molecule, d (Å)	51	50
Calculated molecular weight, M	60,000	40,000

[a] Calculations based on a lipoprotein, density 1.1, would give molecular weights of 48,000 for frog and 32,000 for bovine rhodopsin. The molecular weight estimated for bovine rhodopsin was 35,000–37,000 (Abrahamson and Fager, 1973).

[b] Data taken from Wolken (1961, 1966, 1975).

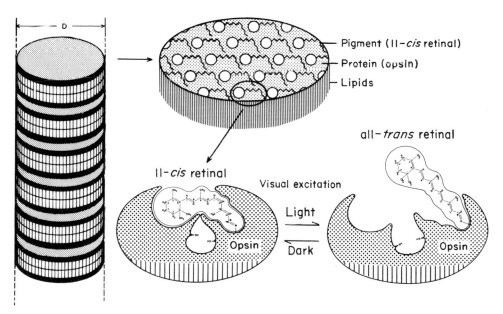

Fig. 9.21 Model for molecular structure of retinal rod, showing the possible molecular geometry of retinal with opsin (rhodopsin) in the membrane of the retinal rod. Note the conformational change upon light absorption that releases the 11-*cis*-retinal from rhodopsin to the all-*trans*-retinal. After Hubbard and Kropf (1959); from Wolken (1975), p. 211.

III. Remarks

We have emphasized that all vertebrate photoreceptors—the retinal rods and cones—consist of lamellae, which are membranes of proteins and lipids separated by equidistant aqueous layers. Associated with the protein of the membrane is the chromophore, retinal. In invertebrates, the photoreceptors—rhabdomeres—are membranes formed into microtubules that show hexagonal or lamellar packing, depending on the angle of cut through the photoreceptors. These various photoreceptor structures, from the invertebrate to the vertebrate, are compared in Fig. 9.22. Most

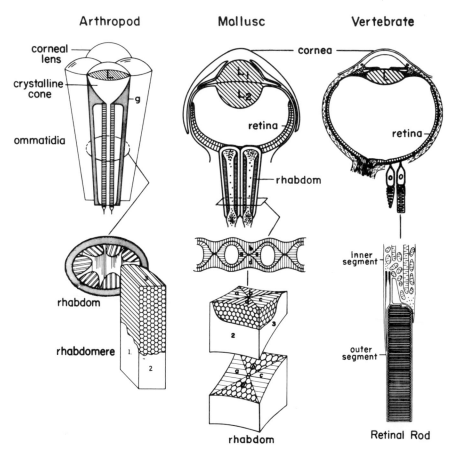

Fig. 9.22 Structural development of the eye, from arthropod compound eyes, to molluscs to the vertebrate eye, showing the optical and photoreceptor structures. From Wolken (1971, p. 142).

important in these studies of the photoreceptors is that their structure and properties are similar to the cell membrane, for it has been suggested that the origin of the photoreceptor lamellae goes back to the cell membrane (Wolken, 1975).

Like that of the cell membrane, the phospholipid in the photoreceptor membranes is fluid and can undergo phase transitions at critical temperatures. Therefore, from these structural and chemical studies, we can consider that the photoreceptors possess properties ascribed to liquid crystals, for the photoreceptors, like liquid crystals, respond to light, to temperature, to pressure, to electrical stimulation, and to changes in the chemical environment. If indeed what we have observed are liquid crystals, then the understanding of the liquid crystalline state becomes important in interpreting biological structures at the molecular level.

REFERENCES

Abrahamson, E. W., and Fager, R. S. (1973). The chemistry of vertebrate and invertebrate visual photoreceptors. *Curr. Top. Bioenerg.* **5**, 125.

Blasie, J. K. (1972). Location of photopigment molecules in the cross-section of frog retinal disk membranes. *Biophys. J.* **12**, 191.

Blaurock, A. E., and Wilkins, M. H. F. (1969). Structure of frog photoreceptor membranes. *Nature (London)* **233**, 906.

Bouligand, Y. (1972). Twisted fibrous arrangements in biological materials and cholesteric mesophases. *Tissue & Cell* **4**, 189.

Bouligand, Y. (1974). Recherches sur les texture des états mesophores. *J. Phys. (Paris)* **35**, 215.

Hubbard, R., and Kropf, A. (1959). Molecular aspects of visual excitation. *Ann. N.Y. Acad. Sci.* **81**, 442.

Lubbock, J. (1882). "Ants, Bees and Wasps." Appleton, New York.

Luzzati, V., Gulik-Krzywicki, T., and Tardieu, A. (1968). Polymorphism of lecithins. *Nature (London)* **218**, 1031.

Miller, W. H., and Bernard, G. D. (1968). Butterfly glow. *J. Ultrastruct. Res.* **24**, 286.

Neville, A. C. (1975). "Biology of the Arthropod Cuticle." Springer-Verlag, Berlin and New York.

Neville, A. C., and Caveney, S. (1969). Scarabaeid (June) beetle exocuticle as an optical analogue of cholesteric liquid crystals. *Biol. Rev. Cambridge Philos. Soc.* **44**, 531.

Rothschild, K. J., Andrew, J. R., De Grip, W. J., and Stanley, H. E. (1976). Opsin structure probed by Raman spectroscopy of photoreceptor membranes. *Science* **191**, 1176.

Schmidt, W. J. (1935). Doppelbrechung, Dichroismus und Feinbau der Aussengleider der Sehzellen vom Frosch *Z. Zellforsch. Mikrosk. Anat.* **22**, 485.

Schmidt, W. J. (1937). "Die Doppelbrechung von Karoplasma, Zytoplasma, und Metaplasma, Protoplasma," Monogr. II. Borntraeger, Berlin.

Schmidt, W. J. (1938). Polarisationsopische Analyse eines Eiweisslipoidsystems erlautert am Aussenglied der Sehzellen. *Kolloid-Z.* **85**, 137.

von Frisch, K. (1949). Die Polarisation des Himmelslichtes als Faktor der Orientieren bei den Tänzen der Bienen. *Experientia* **5**, 397.

von Frisch, K. (1950). "Bees: Their Vision, Chemical Senses and Language." Cornell Univ. Press, Ithaca, New York (revised edition, 1971).

von Frisch, K. (1967). "The Dance Language and Orientation of Bees." Belknap Press, Cambridge, Massachusetts.

Waterman, T. H., Fernandez, H. R., and Goldsmith, T. H. (1969). Dichroism of photosensitive pigment in rhabdoms of the crayfish *Orconectes*. *J. Gen. Physiol.* **54,** 415.

Wolken, J. J. (1961). A structural model for a retinal rod. *In* "The Structure of the Eye" (K. Smelser, ed.), p. 173. Academic Press, New York.

Wolken, J. J. (1966). "Vision: Biochemistry and Biophysics of the Retinal Photoreceptors." Thomas, Springfield, Illinois.

Wolken, J. J. (1971). "Invertebrate Photoreceptors: A Comparative Analysis." Academic Press, New York.

Wolken, J. J. (1975). "Photoprocesses, Photoreceptors and Evolution." Academic Press, New York.

Wolken, J. J. (1977). *Euglena:* The photoreceptor system for phototaxis. *J. Protozool.* **24,** 518.

Chapter 10
Fibrous Protein Structures and Effectors

I. INTRODUCTION

In Chapter 9 considerable attention was focused on the eye—the optical and the photoreceptor structures of invertebrates and vertebrates, including those of man. In invertebrates, the molecules that form the cuticle and the corneal lens of the eye are chemically similar. That is, the major structural molecule is the polysaccharide polymer, chitin. The cuticle and corneal lens in their development form lamellae, but in the formation of these structures, the lamellae from layer to layer twist into a helicoid (Figs. 9.14–9.17). Such a structural development has been likened to that of a liquid crystal. Furthermore, we related such a helical structure of the corneal lens to the polarization of light and hence to function in vision for these animals. This then leads us to examine in more detail the cornea of vertebrate eyes (Fig. 9.1). Before we can do so, it is of interest to discuss the major protein molecule of the cornea, which is collagen.

II. FIBROUS PROTEIN STRUCTURES

A. Collagen

A structural fibrous protein found throughout all animal phyla from coelenterates to mammals, from sponges to man, is collagen. It is a major component of bone, cartilage, connective tissue, skin, and cornea of vertebrate eyes.

Collagen is composed of fibers in the form of bundles. Each fiber, as identified by electron microscopy, has a characteristic periodic structure (Fig. 10.1a). Collagen fibers can be solubilized in 1 M NaCl or dilute acetic acid; if the solution is then diluted with 5% NaCl, the collagen molecules

Collagen

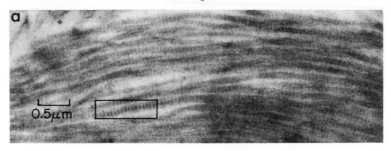

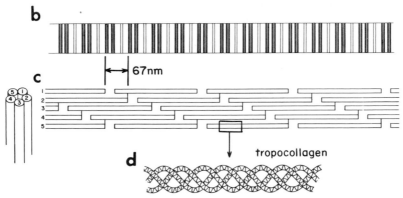

Fig. 10.1   (a) Collagen fibers, in a section through the rat brain. (b) Periodic spacing in stained collagen fibers. Note spacing of 67 nm (in other collagen fibers these range from 64 to 70 nm). (c) Schematic representation to show how the tropocollagen molecules overlap in the collagen fiber in (b). (d) Tropocollagen triple helix structure. These are composed of three left-handed helices which are given a right-hand twist to form the threefold superhelix. (Refer to Dickerson and Geis, 1969.)

reaggregate to form fibers again. The fundamental structural unit of collagen is the *tropocollagen* molecule. The tropocollagen molecules are thought to pack with a displacement of one-quarter of their length to form an overlapping collagen fibrils having a characteristic banding pattern every 67 nm (67–70 nm). These fibrils are about 0.5 μm in diameter, but swell to six or more times in a dilute acid medium and when stained for electron microscopy. The tropocollagen molecule consists of three polypeptide chains that are helically wound around one another to give a triple helix (Fig. 10.lb–d) with a pitch length of one complete turn of about 30 amino acid residues. Each polypeptide chain is composed of hydroxyproline, proline, and glycine. The triple-helix structure of tropocollagen was primarily worked out by Rich and Crick (1961) and Ramachandran (1963) and is unique in its structure for a fibrous protein.

Collagen strength and stability reside in the chemical cross-links between the molecules. Therefore, any structural defects due to cross-linkage of the polypeptide chains of collagen molecules are particularly important in the failure of skeletal and other structural tissue cells (Woodhead-Galloway and Hukins, 1976). These defects are brought about by age, various disease states such as diabetes, and physiological changes which accompany them.

B. The Cornea

The structural molecule of the cornea of the eye is collagen. In the cornea the collagen fibers form lamellae of uniform thickness and appear as stacked plates. The fiber axes in adjacent lamellae make large angles with one another, as in the structure of the corneal lens of compound eyes (Fig. 9.15). Such a structural arrangement of collagen in the cornea, that of a regular lattice, forms the elements for a diffraction grating, to the light scattering characteristics and to its transparency. Therefore, collagen in the cornea of vertebrate eyes has the same structural relationship to that of chitin in the corneal lens of invertebrate eyes (see Fig. 9.14).

III. EFFECTORS: MUSCLE AND NERVE

As organisms evolved, certain cells developed specialized structures to transmit signals from the receptors, the nerve cells; and cells to respond to the signal, the muscle cells. The development of nerve and muscle cells made it possible for the organism to function more efficiently in an integrated manner, as a whole animal.

We have already noted the previous studies of Rinne (1933) on liquid crystals and that of Needham (1950) on the paracrystalline state in living cells. According to Needham, two striking examples of paracrystalline structures are those of striated muscle and myelinated nerve. Let us examine the structures of striated muscle and myelinated nerve to see whether any analogies can be found in their molecular structure to that of liquid crystals.

A. The Muscle Cell

Muscle cells are specialized for the function of contraction—to do work for the organism. Striated muscle consists of special fibers, *myofibrils,*

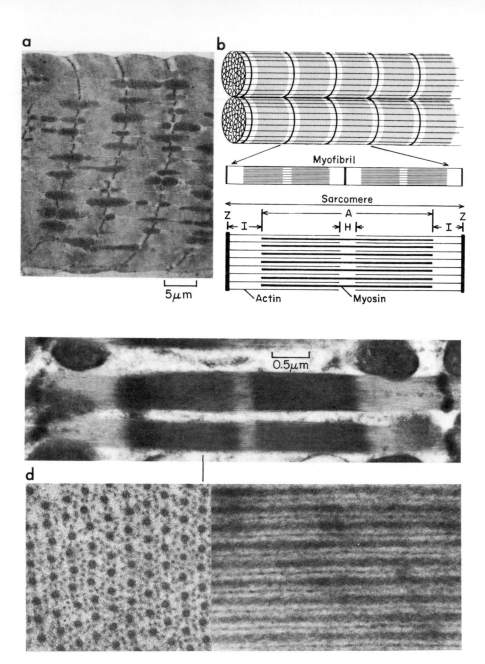

Fig. 10.2 (a) Muscle bundle in the insect (June beetle). Note bands, mitochondria, and lamellar structure. (b) Schematic drawing of two such bundles of myofibrils. Note arrangement of bands (Z line), and lines (I), (A), and (H), also thick filament myosin and thin filament actin. (c) Longitudinal section through several myofibrils. Compare to structures as noted in (b). (d) Cross section through (c) and longitudinal section through (c) at higher magnifications.

running the length of the muscle cell (Fig. 10.2a). Myofibrils are rods about 1 μm in diameter. When examined with the light microscope, myofibrils are seen to have a banded structure. These bands repeat themselves at approximately 2-μm intervals. Each such unit is called a *sarcomere* (Fig. 10.2a,b). Under polarized light the muscle fibers exhibit birefringence. The bands along the fibers are designated by various initials. The dark A band is anisotropic and the I band is isotropic. Within the I band is a Z line and the distance between the Z lines defines the sarcomere (Fig. 10.2c).

Electron microscopy has shown that the myofibril consists of thick and thin filaments. The thick filaments are almost entirely the protein *myosin*. The thin filaments are composed of another protein, *actin* (Fig. 10.2b,c). The muscle contracts when the myosin molecules make contact with the actin molecules.

Actin and myosin can be extracted from muscle tissue. When these two proteins are in solution together, they form the complex *actomyosin*. Albert Szent-Györgi discovered some time ago that the actomyosin complex can be precipitated and from the precipitated complex, muscle fibers can be formed which will contract when immersed in an ATP solution. Chemically, the contractile structure of muscle consists almost entirely of three proteins: myosin, actin, and tropomyosin. About half of the dry weight of these contractile proteins is myosin. Myosin is also the enzyme which catalyzes the removal of a phosphate group from ATP. The energy liberating reaction is an event associated with the contraction.

The muscle system is not as simple as we have described (Fig. 10.2). The signal that initiates the event of muscle contraction is the release of calcium ions from the membrane-bound storage sites by the arrival at the muscle of a nerve impulse. The signal is detected and acted on by two additional proteins, *tropomyosin* and *troponin,* which are positioned along the actin filament. It is believed according to Cohen (1975) that the calcium ions bind to troponin, which then modifies the positions of the tropomyosin molecule, so that myosin can make contact with the actin molecules. However, the mechanism by which the thin filament actin is pulled past the thick filament myosin is not yet completely understood. The elucidation of the structure of striated muscle is primarily due to the extensive studies of Huxley (1973) and his associates over the past two decades.

The similarity of striated muscle to that of a smectic liquid crystal structure was suggested by Needham (1950). More recently, April (1975a,b) has put forth the idea that the myofilament lattice of striated muscle can exist in more than one liquid crystalline state. This was based on X-ray diffraction studies of the muscle fibers of the walking legs of the crayfish

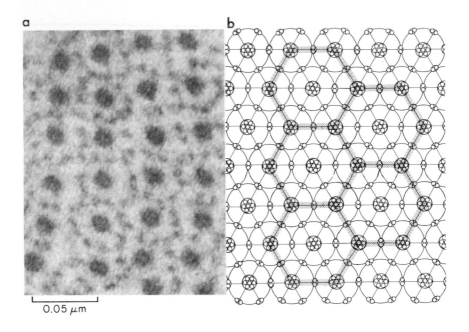

0.05 μm

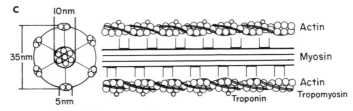

Fig. 10.3 (a) Cross section of myofibril, as seen from the end. The thick filament, myosin, and thin filament, actin, lie beside one another in a hexagonal array (June beetle). (b) Schematic drawing of how the actin and myosin are arranged, as visualized from the electron micrograph (a). (c) Schematic drawing of the longitudinal section, showing actin associated with tropomyosin and myosin. These are the three major proteins of muscle myofibrils.

(*Orconectes*), which supported his suggestion that the resting myofilaments form a smectinlike liquid crystalline packing of molecules. The liquid crystalline lattice of striated muscle would appear to be a necessary adaptation whereby the lattice provides rigidity for the transmission of contractile forces and for the anchoring of the flexible cross bridges, which at the same time enables shortening to occur by the mechanism of sliding filaments (Fig. 10.3).

The interesting observations from the electron micrographs of the June beetle muscle were those in cross section and longitudinal sections through the A band (Fig. 10.2d and 10.3a) the spatial arrangement of actin and myosin can be seen. In Fig. 10.3a the actin filament appears as six doublets surrounding myosin, and myosin as six filaments with probably an additional central filament. In cross section they are hexagonally packed structures as described by Huxley (1973). We have schematized this structure in Fig. 10.3b from our electron micrographs.

Outside of muscle, contractile proteins are found in flagella and cilia (see Figs. 7.14 and 7.15), which are associated with the cell's motility and have similarities to the structure of muscle (compare Fig. 7.15 to 10.3). Muscle proteins have also been isolated from fungi, protozoa, algae, and from the cell walls of plants which bend or contract upon mechanical stimulation.

We see then that the muscle system is more universal throughout the plant and animal phyla than originally thought. There is a relationship between the structures designed for movement to that of the highly specialized muscle cells—and even to that of liquid crystals.

B. The Nerve Cell

The nerve cell, or neuron, consists of a cell body. The cell body contains a nucleus, mitochondria, Golgi, ribosomes, endoplasmic reticulum, and microtubules (see schematized cell, Fig. 7.2). Dendrites, dendritic processes, extend from the cell body. The cell body receives messages from other nerves which make contact with its dendrites. From the cell body emerges a nerve fiber, or *axon* (Fig. 10.4). Structurally the axon is a cylinder with a conducting core and a surface membrane of relatively high resistance. The nerve signal passes along the axon and down the axon terminal and makes a contact called a synapse. At the terminal the electrical signal is converted into a chemical message in the form of a neurotransmitter molecule (e.g., acetylcholine), which is discharged from vesicles in the terminals and flows on to the membrane of the next neuron. For a more complete description of the various types of nerve cells and how they function, reference should be made to the works of Hodgkin (1964) and Katz (1966).

Because of their axonal processes, nerve cells have a great need for structural support. The nerve axon does not possess the highly ordered arrangement as in muscle (Fig. 10.2).

An interesting structural aspect is that vesicles and microtubules which contain phospholipids are observed. The nerve membrane is an important

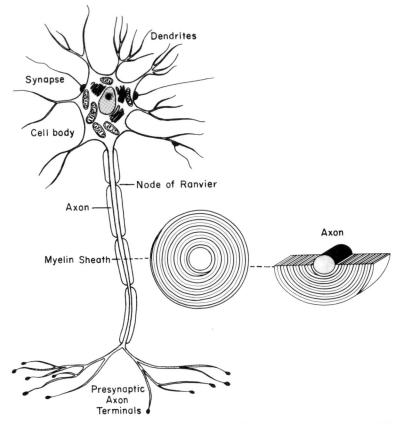

Fig. 10.4 Schematic drawing of a typical myelinated nerve cell. Note cell body, axon, and myelinated axon.

part of the nerve cell and is intimately connected with the transmission of signals.

In myelinated nerve fibers, the axons are wrapped in a myelin sheath, membranes that surround the axon (Fig. 10.4 and 10.6). The myelin sheath has a strong birefringence and is a highly ordered structure. It is formed by a bimolecular layer of lipid oriented radially and with concentric layers of protein (Fig. 10.5, also see Fig. 7.5a, 7.9). In the autonomic nervous system most nerve fibers are unmyelinated and are contained within invaginations of the plasma cell membrane of Schwann cells (Fig. 10.7)

Among the invertebrates, there are very large unmyelinated axons. One such axon was discovered by Young (1936) in the squid, *Loligo,* and is referred to as the giant axon. The squid giant axon behaves like a single

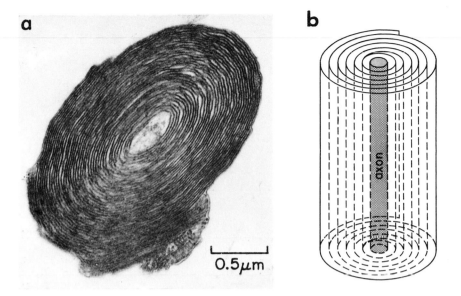

Fig. 10.5 (a) Myelinated nerve in the frog retina. (b) Schematic drawing of myelin sheath surrounding axon, depicting the layers of lipid and protein. From Wolken (1966). Reprinted by permission.

cell. When viewed under the light microscope, the giant axon appears as a transparent cylinder of cytoplasm surrounded by a thin sheet of connective tissue. In a freshly isolated giant axon the cytoplasm appears to be a gel and exhibits properties of both a liquid and a solid. The cytoplasm of the giant axon is weakly birefringent and its structure is likened to that of axially oriented protein micelle.

IV. REMARKS

What becomes apparent in these investigations of biological structures is that they all exhibit some kind of order—as observed in the numerous figures through the text. They were of course influenced by the methods used to visualize their molecular structure. These methods are primarily microscopy, polarization microscopy, and electron microscopy, as well as optical diffraction, X-ray diffraction, and spectroscopy.

There are three basic configurations of the polypeptide chain in fibrous proteins: the α-helix (Fig. 5.2 and 7.15), the triple helix, just described for collagen (Fig. 10.1), and the β-pleated sheet as in silk and other proteins.

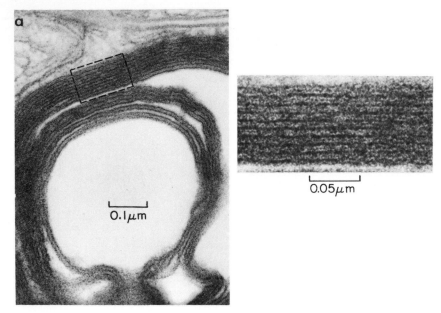

Fig. 10.6 (a) Myelinated axon in the eye of the crustacean copopod *Copillia*. (b) Enlarged area of (a) (cf. Figs. 10.4 and 10.5). From Wolken (1975).

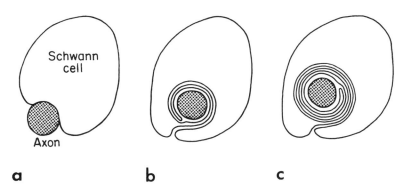

Fig. 10.7 Schematic drawing showing the process of myelination. Invagination of nerve axon in Schwann cell (a), followed by wrapping of successive layers of Schwann cell membranes around axon (b,c).

By adapting these various forms of molecular architecture biological structures of the most diverse properties and function are constructed.

We have described these biological structures in the simplest terms, for they are exceedingly complex. Much more is known about these fiber proteins than is presented here and reference should be made to those studies cited in the bibliography and to other recent researchers. It is evident that considerably more work is required to understand how collagen, muscle, and nerve are organized and function at the molecular level. Our bias, of course, was directed to finding a rational explanation for these biological structures which exhibit in their molecular organization the properties of the liquid crystalline state.

REFERENCES

April, E. W. (1975a). Liquid-crystalline characteristics of the thick filament lattice of striated muscle. *Nature (London)* **257**, 139.
April, E. W. (1975b). Myofilament lattice: Studies on isolated fibers. *J. Mechanochem. Cell Motil.* **3**, 111.
Cohen, C. (1975). The protein switch of muscle contraction. *Sci. Am.* **233**, 36.
Dickerson, R. E., and Geis, I. (1969). "The Structure and Action of Proteins." Harper, New York.
Hodgkin, A. L. (1964). "The Conduction of Nervous Impulses." Thomas, Springfield, Illinois.
Huxley, H. E. (1973). Structural changes during muscle contraction. *Cold Spring Harbor Symp. Quant. Biol.* **37**, 361.
Katz, B. (1966). "Nerve, Muscle and Synapse." McGraw-Hill, New York.
Needham, J. (1950). "Biochemistry and Morphogenesis," p. 661. Cambridge Univ. Press, London and New York.
Ramachandran, G. N., ed. (1963). "Aspects of Protein Structure," p. 39. Academic Press, New York.
Rich, A., and Crick, F. H. C. (1961). *J. Mol. Bio.* **3**, 483.
Rinne, F. (1933). Investigations and considerations concerning paracrystallinity. *Trans. Faraday Soc.* **29**, 1016.
Wolken, J. J. (1966). "Vision: Biochemistry and Biophysics of the Retinal Photoreceptors," p. 59. Thomas, Springfield, Illinois.
Wolken, J. J. (1975). "Photoprocesses, Photoreceptors and Evolution," p. 59. Academic Press, New York.
Woodhead-Galloway, J., and Hukins, W. L. (1976). Molecular biology of cartilage. *Endeavour* **35**, 36.
Young, J. Z. (1936). Structure of nerve fibres and synapses in some invertebrates. *Cold Spring Harbor Symp. Quant. Biol.* **4**, 1.

Chapter 11
Membranes

I. INTRODUCTION

We have emphasized the importance of the cell and cellular membranes to the functioning of various types of cells because our understanding of cellular processes is directly related to the molecular structure of these cellular membranes.

II. MOLECULAR STRUCTURE

The molecular structure of cell membranes, their molecules, properties, and behavior, are described in some detail by Bretscher (1973), Chapman (1973), Guidotti (1972), Quinn (1976), and Stein (1967), as well as in other references noted herein. Membranes are complicated structures and it is not our purpose to review all that is known about cell membranes. The question we have asked, though, is, Is the cell membrane a liquid crystalline system? We have stressed such analogies in exploring the various cellular structures, but let us examine more closely the properties and behavior of lipids and proteins of which the membranes are constructed.

The major structural molecules of cell membranes are lipids and phospholipids (Fig. 5.5 and Table 7.1), with specific proteins (Tables 9.1 and 11.1). It will be noted in Table 11.1 that there are differences in the total lipids to the total proteins and that some approach a 50:50 ratio. To know how the protein molecules are structured and oriented in relation to the molecular structure of the lipids in membranes is a necessary requisite toward understanding how they function.

Experimentally, membranes can be formed with phospholipids. Proteins, enzymes, and pigments can be incorporated into the membrane. These membranes have been studied by microscopy, electron micros-

TABLE 11.1

Composition of Membranes[a]

	Lipid (%)	Protein (%)
Red blood cell (bovine)	30–40	60–70
Red blood cell (human)	40	60
Myelin	75	25
Myelin sheath (nerve)	45–60	20–40
Ox brain	18–23	73–78
Mitochondrion membranes	20–50	50–60
Mitochondrion (beef heart inner membranes)	25	75
Chloroplast lamellae	50	50

[a] Data taken in part from Haggis *et al.* (1965) and Branton and Park (1968).

copy, X-ray diffraction, and various techniques in spectroscopy. Considerable data have been obtained about the structure and behavior of these membranes, as well as from isolated cell membranes, and membranes *in situ* by these methods.

Phospholipids swell in water and form many spherical bodies composed of concentric layers, lamellae, with water trapped between them. (Figs. 5.6 and 5.7). When such phospholipid bodies are sonicated they form into vesicles referred to as liposomes (Bangham, 1968) that are about 250 Å in diameter; the lipid layers are of the order of 50 Å thick. Both phospholipid structures formed in an aqueous environment and liposomes are liquid crystalline systems and they resemble myelin structures in the cell (Fig. 7.9). Readers should refer to a recent symposium on "Liposomes" (Papahadjopoulos, 1978).

Phospholipids also depend very markedly on temperature; on heating they undergo an endothermic transition at a temperature well below the true melting point, which would be at about 200°C. At this transition temperature, a change of state occurs from the crystalline, or gel, to the liquid crystalline state—which is associated with increased conformational freedom for the lipid fatty acid chains. The transition temperature increases with increasing length of the fatty acid chain, and decreases with increasing unsaturation in the chain (Lee, 1975).

Below the transition temperature, in the gel (crystalline) phase, the phospholipids adopt a bilayer structure, in which the fatty acid chains are packed in ordered hexagonal arrays where motion of the fatty acid chains is highly anisotropic and restricted. At the transition temperature, there is

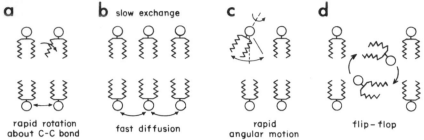

Fig. 11.1 Mobility of the phospholipid hydrocarbon chain. (a) Rapid rotation about C–C bond in the hydrocarbon. (b) Lateral diffusion in the plane of the membrane. (c) Angular motion of the phospholipid molecule. (d) Flip-flop of the phospholipid molecules across the bilayer.

a 50% increase in the surface area occupied by the lipid and appreciable motion becomes possible about the C–C bonds of the fatty acid. The motion about the C–C bonds in lipid bilayers has been studied by ^{13}C NMR, and ^{13}C spectra show resonances for many of the fatty acid chain carbons. It was found that there is a very marked motional gradient within the phospholipid molecule in the liquid crystalline phase. The motion about the C–C bonds within the fatty acid ^{13}C NMR data provided evidence suggestive of an axial rotation of the whole lipid molecule in the plane of the bilayer. We have schematized the degrees of freedom possible for the phospholipid chains in Fig. 11.1.

One can say that a membrane containing phospholipids with little unsaturation is less fluid than one with much unsaturation. The control of fluidity of the components of cell membranes may be related to the diffusional characteristics of molecules and ions passing in and out of the membrane. The state of the phospholipid in a membrane—in a gel or in a liquid crystalline state—can be expected to have a marked effect on the function of the membrane. Thus, small molecules will be able to move relatively easily through a membrane in which the phospholipids are in a liquid crystalline state, but not one in which they are in the gel state.

Membrane proteins are not well characterized nor is their molecular structure precisely known, except for a few membranes. Generally, membrane proteins can be either extrinsic (peripheral) or intrinsic. The extrinsic proteins can be removed from the membrane by aqueous extraction, that is, by changing the ionic strength and/or the pH. Intrinsic proteins can only be removed from the membrane by the action of detergents, surfactants, or organic solvents which solubilize the lipids. The extrinsic proteins are associated with membrane surfaces and are bound mostly by electrostatic forces. The intrinsic proteins are found in the interior of the membrane and are held primarily by van der Waals forces.

Low molecular weight proteins with water can generate their own liquid crystalline structures, especially if the protein possesses a high degree of polarity like that of the phospholipids. The polar part of the protein molecule is then water soluble, whereas the organic part of the molecule is soluble in the organic lipids of the cell membrane.

III. MEMBRANE MODELS

Since the Danielli and Davson (1935) membrane model was introduced, various modifications to the lipid bilayer have been proposed. Vanderkooi and Green (1971) suggested that some membrane proteins might not be electrostatic, but bimodal, that is, their protein molecules would possess both polar and nonpolar groups like the phospholipids. Proteins in a globular configuration would fit directly into a lipid bilayer, their hydrocarbon chains and their polar groups bonded to lipid heads.

Another structural model is that described by Singer and Nicholson (1972); the model has many appealing properties that a membrane should possess. In this model the protein can float in the lipid, because lipids and proteins are mobile when they are in a fluid (or melted) state. Also, both fluid and solid regions may be present in the same membrane, and Oldfield (1973) indicates that it is possible for membranes to contain some of their lipids in a crystalline state. These structural models for cellular membranes have been reviewed by Bretscher (1973), Capaldi et al. (1973), Eisenburg and McLaughlin (1976), and Meyers and Burger (1977).

We have schematized these various membrane models in Fig. 11.2. In Fig. 11.2a the proteins are on the surfaces of the lipid layers. In Fig. 11.2b another protein can extend into the lipid layer and is capable of moving through the lipid bilayer by rotational and lateral diffusion (Fig. 11.2c). In Fig. 11.2d a protein molecule can extend through the entire width of the lipid bilayer. In Fig. 11.2e the proteins reside on the surfaces of the lipid layers, but another protein lies between the lipid layers. In Fig. 11.2f this model is further expanded to show that another protein can wrap around the surface proteins in an α-helix, as illustrated for the striated muscle filaments (Fig. 10.3). These models indicate that a number of possibilities exist for proteins to be associated with other proteins on the surfaces of the lipid layers, in the lipid layers, and between the lipid bilayer.

An example of a highly specialized membrane is that of the bacterium *Halobacterium halobium,* which grows in a high salt concentration (25% NaCl) at temperatures near 44°C, and in direct sunlight. Their cell membrane is a complex system in which the visual pigment rhodopsin resides in its cell "purple membrane." A model for the membrane constructed

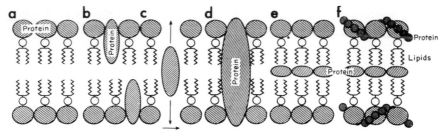

Fig. 11.2 Membrane models. (a) The lipid bilayer with associated proteins. (b) The lipid bilayer in which proteins are not only on the surfaces, but another protein lies between the bilayer. (c) The protein with rotational reaction as well as lateral diffusion. (d) A membrane in which a protein extends above and beyond the lipid bilayer. (e) The lipid bilayer in which the protein not only on the surface, but another protein lies between the lipid bilayer. (f) A membrane similar to (e) but another protein is wrapped in an α-helix around the surface protein.

by Unwin and Henderson (1975) indicates that the protein molecules comprise seven rod segments as α-helices, which are oriented perpendicular to the plane of the membrane. In a unit cell, three helices from each of the three protein molecules form a nine-membered inner ring which is surrounded by a 12-membered outer ring of the remaining helices. The spaces between neighboring rings are filled by the lipid bilayer molecules.

IV. REMARKS

The lipid bilayer is the basic structural unit for cell membranes. Phospholipids are the major molecules in the lipid bilayers (Fig. 5.5 and Table 7.1). Phospholipids in water form liquid crystals and resemble myelin structures found in cells. Therefore, the molecular packing of the lipids dictate the skeletal structure of the membrane. Proteins and enzymes are molecularly associated with these lipid bilayers.

Although there is no one membrane model that applies to all biological membranes, we have schematized the various cell membranes (Fig. 11.2) which best fit the experimental data at our present level of understanding.

Membranes are not static structures but exist in a dynamic state, and their molecules have mobility. Labeling experiments show that the membrane components do not exchange rapidly from one side of the bilayer to the other, but most other motions are possible. It appears that the membrane has a fluid lipid matrix in which embedded molecules can move rather freely.

The dynamic nature of the membrane is observed, for example, in that the fastest molecular motion is about the C–C bonds in the lipid hydrocarbon. The lipids can exhibit translational motion and can undergo fast lateral diffusion within the plane of the membrane (Fig. 11.1). Although lateral diffusion of lipids is known to occur rapidly, different lipid orientation in different regions of a membrane is possible (Bergelson and Barsukov, 1977). Proteins show comparable motions but are slower. The membrane also has static features, for not all membranes exhibit highly fluid properties; the degree of fluidity will depend on the lipid composition. In the boundary layers where the lipid is dissolved in water the structure is rather immobile. The protein can also be mobile, but their motional properties in the membrane will depend on the lipids which surround a specific protein and the degree of interaction between the lipid and other proteins in the membrane.

We have examined the structure of the cell membrane in a variety of cells (Figs. 7.8 and 7.9) and the cellular membranes of the chloroplast (Figs. 8.2–8.5), the mitochondrion (Figs. 7.11 and 7.12), the striated muscle (Figs. 10.2 and 10.3), myelinated nerve (Figs. 10.4–10.7) and the retinal rods of vertebrate eyes (Figs. 9.2–9.4), and the rhabdoms of invertebrate eyes (Figs. 9.8–9.11) that differ for function, but there is a similarity in that they possess certain structural and behavioral properties in common.

For the membrane to function, we have concluded that the cell and cellular membranes are liquid crystalline structures and that such a structure permits ion transfer and the movement of molecules, necessary requirements for all cell membranes.

REFERENCES

Bangham, A. D. (1968). *Prog. Biophys. Mol. Biol.* **18**, 29–95.

Bangham, A. D. (1978). Uses of lipid vesicles: An overview. *In* Liposomes and their uses in biology and medicine. *Ann. New York Acad. Sci.* **308**, 2–7.

Bergelson, L. D., and Barsukov, L. I. (1977). Topological asymmetry of phospholipids in membranes. *Science* **197**, 224.

Branton, D., and Park, R. B., eds. (1968). "Biological Membrane Structure," pp. 13 and 14. Little, Brown, Boston, Massachusetts.

Bretscher, M. S. (1973). Membrane structures: Some general principles. *Science* **181**, 622.

Capaldi, R. A., Komap, H., and Hunter, D. R. (1973). Isolation of a major hydrophobic protein of mitochondrial inner membrane. *Biochem. Biophys. Res. Commun.* **55**, 655.

Chapman, D. (1973). Some recent studies of lipids, lipid-cholesterol and membrane systems. *Biol. Membr.* **2**, 91.

Danielli, J. F., and Davson, H. (1935). A contribution to the theory of permeability of thin films. *J. Cell. Comp. Physiol.* **5**, 495.

Eisenburg, M., and McLaughlin, S. (1976). Lipid bilayers as models of biological membranes. *BioScience* **26**, 436.

Guidotti, G. (1972). Membrane proteins. *Annu. Rev. Biochem.* **41**, 131.

Haggis, G. H., Michie, D., Muir, A. R., Roberts, K. B., and Walker, P. M. B. (1965). "Introduction to Molecular Biology," p. 161. Wiley, New York.

Lee, A. G. (1975). Interactions within biological membranes. *Endeavour* **34,** 67.

Meyers, D. I., and Burger, M. M. (1977). Puzzling role of cell surfaces. *Chemistry* **50,** 36.

Oldfield, E. (1973). Are cell membranes fluid? *Science* **180,** 982.

Papahadjopoulos, D., ed. (1978). Liposomes and their uses in biology and medicine. *Ann. New York Acad. Sci.* **308,** 1–462.

Quinn, P. J. (1976). "The Molecular Biology of Cell Membranes." Univ. Park Press, Baltimore, Maryland.

Singer, S. J., and Nicholson, G. L. (1972). The fluid mosaic model of the structure of cell membranes. *Science* **175,** 720.

Stein, W. D. (1967). "The Movement of Molecules across Cell Membranes." Academic Press, New York.

Unwin, P. N. T., and Henderson, R. (1975). Molecular structure determination by electron microscopy of unstained crystalline specimens. *J. Mol. Biol.* **94,** 425.

Vanderkooi, G., and Green, D. E. (1971). New insights into biological membrane structure. *BioScience* **21,** 409.

Applications to Medicine
and
Summary

Chapter 12
Liquid Crystals in Medicine

I. INTRODUCTION

We have considered liquid crystals in a variety of living cellular structures and have particularly emphasized cellular membranes throughout our discussions. However, one of the first applications of liquid crystals to biological systems was as an analytical tool to measure temperature changes. The term *thermography* is now commonly used to identify any diagnostic technique involving a temperature gradient. Liquid crystals can also be considered to play some role in various disease states of man, for example, in the hardening of the arteries, in gallstone formation, and in sickle cell anemia.

II. TEMPERATURE SENSING

A. General Discussion

The type of liquid crystal used in these measurements is the cholesteric–nematic structure which changes color as the temperature changes. Individual chiral compounds and mixtures of chiral compounds (e.g., cholesteryl esters) have different liquid crystalline ranges, and over these ranges color changes will take place with temperature changes. As the temperature increases, the color changes from brilliant reds to yellows, to greens, to blues, to violets, with the color finally disappearing as the isotropic liquid appears. On cooling the color changes are reversible. Thermography (thermal mapping) relies on the conductive heat exchange between the sensing liquid crystal film and the test surface.

Some calibrated mixtures are commercially available. We list a few representative compositions in Table 12.1, covering the temperature range 0°–250°C.

Table 12.2 lists typical mixtures covering the temperature range 30°–

TABLE 12.1

Temperature Intervals of Color Bands of Mixtures of Cholesteryl Esters[a]

Components: cholesteryl	Ratio	Temperature range (°C)
Oleyl carbonate, acetate	80:20	0–4
Oleyl carbonate, phenyl carbonate	80:20	14–16
Oleyl carbonate, acetate	95:5	16–18
Oleyl carbonate, nonanoate, benzoate	65:25:10	17–23
Oleyl carbonate, nonanoate, benzoate	70:10:20	20–25
Methyl carbonate, nonanoate	20:80	22–47
Nonanoate, oleate, crotonate	25:55:20	22–25
Nonanoate, oleate, crotonate	10:70:20	24–26
Oleyl carbonate, nonanoate, benzoate	45:45:10	26.5–30.5
Oleyl carbonate, nonanoate, benzoate	43:47:10	29–32
Oleyl carbonate, nonanoate, benzoate	44:46:10	30–33
Oleyl carbonate, nonanoate, benzoate	38:52:10	33–36
Oleyl carbonate, nonanoate, benzoate	32:58:10	36–39
Nonanoate, oleate, crotonate	30:60:10	40–42
Nonanoate, propionate	80:20	45–65
Nonanoate, butyrate	80:20	55–57
3-Phenylpropionate, nonanoate	20:80	64–67
Cinnamate, nonanoate	90:10	140–250

[a] Modified from Elser and Ennulat (1976, p. 135).

TABLE 12.2

Temperature Ranges of Color Bands of Mixtures of Cholesteryl Esters[a]

Temperature range (°C)	Wt % cholesteryl		
	Oleyl carbonate	Nonanoate	Benzoate
30–33	44	46	10
31–34	42	48	10
32–35	40	50	10
33–36	38	52	10
34–37	36	54	10
35–38	34	56	10
36–39	32	58	10
37–40	30	60	10

[a] Modified from Elser and Ennulat (1976, p. 73).

40°C with each "recipe" having a color band of 3°C. You will notice that the components are the same in all mixtures, but their ratios are different. Different compositions give color patterns over different temperature ranges. This is a very useful property in diagnostic procedures.

Temperature sensing using cholesteric–nematic liquid crystals has advantages over many other methods. One of these advantages is that a color pattern can be developed without regard to the thickness of the film. Reflected light from a cholesteric film is circularly polarized, either left or right, depending on the compound. The color display is a bulk property with the helix generated at the surface. This property and the fact that the selective reflection is independent of whether the illuminating light is polarized or not give cholesteric films their characteristic iridescence. The cholesteric–nematic liquid crystal gives a selective reflection; the light reflected is complementary to the transmitted light. If ordinary light is divided into two beams, one circularly polarized in one sense will be transmitted and the other polarized in the opposite sense is totally reflected. No reversal of the sense of rotation occurs on reflection. In other words, if left-handed circularly polarized light is reflected from a left-handed cholesteric film it remains polarized to the left.

The color of the reflected light is dependent on (1) the temperature of the liquid crystal film, (2) the angle of incidence of the light illuminating the film, (3) the viewing angle of the reflected light, and (4) the nature of the compound.

One of the principal advantages of liquid crystals is their ability to measure surface temperatures accurately and to map out thermal regions. This same area may require several thermocouples to accomplish the same area analysis. If liquid crystal films are exposed to the atmosphere, they will decompose slowly. However, when they are encapsulated in agar-agar, gelatin, and other polymeric materials, their lifetime is indefinite.

B. Applications

As a "chemical," thermometer liquid crystals represent the first major breakthrough in nonelectronic temperature measurement in over 150 years. Disposable oral thermometers are commercially available. As temperature sensors, liquid crystals are more legible than many temperature indicators. In addition to legibility, one can have flexibility in the display, since it may be a digit, word, or symbol. The size of the display can also be altered to fit the defined use.

Liquid crystals as temperature sensors are reasonably accurate. One can expect an accuracy of approximately 0.3°C, which is quite acceptable for most medical applications. One can avoid the visual misinterpretations often encountered with conventional devices (for example, the parallex problem). The response time of the liquid crystal is "instantaneous" while an alcohol or mercury thermometer may require several minutes.

The medical applications of cholesteric–nematic liquid crystals include oral thermometry, cutaneous thermography, gynecology, neurology, oncology, pediatrics, surgery, podiatry, and dentistry. To be more specific, these medical applications include detection of breast cancer, location of the placenta, blood flow patterns in extremities of the human anatomy, and observation of skin temperature changes following blockage of the sympathetic nervous system. The observation of skin temperature changes following blockage of the nerve system enables the physician to determine if neurological and vascular pathways are open. Continuous monitoring of the skin temperature over an extended area provides the physician with a detailed and more easily interpreted indication of circulatory patterns than point measurements with thermocouples and thermistors.

A tumor is generally warmer than the surrounding tissue, which makes it easy to identify its contours and thus guide a surgeon where to cut to extract the tumor and leave as much healthy tissues as possible to enhance healing.

The placenta is warmer than the surrounding tissue, and knowing its location aids the obstetrician in determining if the baby can be expected to have normal delivery or whether a Caeserian is in order.

Third degree burns are cooler than second degree burns, because the blood circulation in the third degree area is less than in a second degree burn. Use of liquid crystals to identify the extent of a burn can speed up treatment and avoid the delay of several days for the development of a scab.

C. Detection of Chemicals

1. Color Change

Many chemicals that diffuse into a cholesteric–nematic liquid crystalline film affect the pitch of the helix, which then changes the wavelength of light diffracted and thus the color of the reflected light. Different chemicals (vapors) show different colors for different cholesteric–nematic liquid crystals or mixtures of cholesterics. If the contaminant is dissolved in the cholesteric–nematic liquid crystal, the color change is reversible on evaporation of the solute. If the solute reacts with the liquid crystal the color change is permanent.

The sensitivity of this measurement is in parts per million. This level of sensitivity is often a problem because the film may be activated by contaminants in the atmosphere. Also, lack of selectivity is often frustrating. For example, two different alcohols might trigger the same color response under comparable conditions.

The film as commonly prepared may be affected by oxidation, hydrolysis, and sunlight. Encapsulated films of cholesteric-nematic liquid crystals are more stable than the film that is directly applied. The encapsulated systems, because of the protective cover, are generally not useful in vapor detection.

2. Sense of Smell

The sense of smell very likely employs the liquid crystalline structure. The packing of the molecules in the nasal passages is evidently lamellar or hexagonal and the sense of smell is stimulated by the intrusion of molecules into the structure (membrane). Different molecules will give different disruptive effects of the molecules in the membranes. The sense of smell in humans can be explained in a general way by the penetration of the molecules to be detected into the ordered liquid crystalline structure. The sense of smell can be "flooded" by too much of a given compound and/or too complex a mixture.

III. DISEASES

In the sections which follow we shall consider a number of diseases in which liquid crystals are involved in the diagnosis. In addition to hardening of the arteries, gallstone formation, and sickle cell anemia, we shall name a few rather rare diseases that involve liquid crystal applications and comment on liquid crystals with regard to the processes of ageing and cancer.

A. Hardening of the Arteries

In the laboratory it is well known that phospholipids such as lecithin and sphingomyelin swell in water to give a lamellar, liquid crystalline structure. A schematic representation of molecular packing in this bilayer structure is given in Chapter 3. These water–lipid compositions are sensitive to temperature changes and will undergo order–disorder transitions as the temperature is lowered. Phospholipids from living systems such as brain tissues and red cell membranes interact with water to give liquid crystalline structures. Cholesterol is insoluble in water, and in living sys-

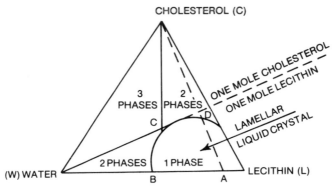

Fig. 12.1 Schematic sketch of a cholesterol–lecithin–water ternary system. One phase is homogeneous, the lamellar liquid crystalline phase. Note along the water–cholesterol side of the triangle cholesterol is totally insoluble in water. Along the lecithin–water axis one sees that lecithin and water form the lamellar liquid crystal between B and A. With more water, the liquid crystalline phase and water are in equilibrium; optically they exhibit anisotropic droplets. Along the cholesterol–lecithin axis the two phase area is the liquid crystal and cholesterol. The three phase area is water–cholesterol–lecithin. (This diagram is drawn to illustrate the phases and is not presented with exact data.)

tems it can precipitate as cholesterol monohydrate to form, for example, gallstones and arteriosclerotic lesions. Cholesteryl esters behave in a comparable way. However, if one takes a three-component system of lecithin, cholesterol, and water, the cholesterol will be solubilized by the water and lecithin to form a liquid crystalline system. The compositions can be as high as a mole of lecithin to a mole of cholesterol. With higher concentrations of cholesterol, it precipitates as cholesterol monohydrate. A ternary system of cholesterol, lecithin, and water can be represented by a general phase diagram (Fig. 12.1). A phase diagram of four components (e.g., cholesterol, cholesteryl esters, lecithin, and water) represents a system which to some degree mocks nature in hardening of the arteries.

Chemical analyses of the deposits on artery walls can give some guidelines on the chemicals that are involved in hardening of the arteries. Analyses of the deposits of a number of patients give results that are helpful but not necessarily accurate. The figures recorded in the tables are given as guides and not as quantitative values. Note that the cholesteryl esters make up most of the deposit in the inclusions. The next question one might ask is what esters are involved. Data on the composition of the esters are recorded in Table 12.4.

Even though the figures in Tables 12.3 and 12.4 are only approximate, they do indicate that cholesteric esters are important chemicals in the body.

TABLE 12.3

Various Lipids in Total Lipid of Inclusions and
Residue in Arterial Deposits[a]

	Total lipid (%)	
Lipid	Inclusion area	Residue area
Cholesteryl esters	95	39
Free cholesterol	2	18
Phospholipids	1	39
Triglycerides	2	4

[a] Modified from Hata and Insull (1973). These data are approximate. For more accurate data, see original article.

Many of the recent studies on arteriosclerotic lesions have been concerned with (1) the characterization and metabolism of specific serum lipoproteins that transport cholesterol and its esters, and (2) the histology, chemical composition, biochemistry, and metabolism of the cells and the chemical composition of the arterial walls.

Through a better understanding of liquid crystals we can now look at the physical state of lipids, cholesterol, and certain cholesteryl esters in living cells. For example, the arteriosclerotic lesions contain large quantities of lipids (e.g., cholesterol, cholesteryl esters, and phospholipids) and the rate of exchange of cholesterol among the plaques, the liquid crystalline phase, and the true solution is slow, indicating that a major fraction of the lipids in the lesions are thermodynamically stable. This is assuming, of course, that there is an equilibrium of the lipids among the plaques, the liquid crystalline state, and the solution. Phase diagrams of the lipids

TABLE 12.4

Various Fatty Acid Esters in Artery Deposits[a]

			Total fatty acid esters in deposit (%)	
Fatty acid	No. of carbon atoms	No. of double bonds	Inclusive	Residue
Myristic	14	0	1	1
Palmitic	16	0	9	11
Stearic	18	0	2	2
Oleic	18	1	50	38
Linoleic	18	2	15	28

[a] Modified from Hata and Insull (1973). These data are approximate. For more accurate data, see original article.

found in arteriosclerotic lesions have been studied, and the compositional limits and the structure of each phase have been determined by standard experimental methods (e.g., microscopy, X-ray, differential thermal analysis). Synthetic systems of lipids showing the compositions similar to those found in lesions allow one to predict the number and the physical state of the lipid systems. That lipid phases exist in intimal lesions has been verified by examining fresh human arteriosclerotic lesions.

Whether arteriosclerotic lesions can undergo reversibility is not really known. In the early changes of the intimal lipid composition, it seems reasonable to predict that the process of ageing of the arterial deposit can be reversed chemically. Solubilization of the excess cholesterol can be accomplished chemically by adding more phospholipid to "dissolve" the free cholesterol. Use of such chemicals as oleic acid are known in the laboratory to solubilize the arterial deposit. An increase in body temperature of a couple of degrees for a period of time might solubilize the arterial deposit.

The molecular geometry of the molecules undoubtedly influence the behavior of the esters. For example, the branching and degree of unsaturation of the acid unit in the ester can modify such properties as the solubility of the esters, the stability of the arterial deposit, and the transition temperatures related to phase changes of the esters in the arterial deposit.

B. Liquid Crystals in Human Bile Fluids

The major components in human bile fluid are water, phospholipids (e.g., lecithin), cholesterol, and bile salts (e.g., sodium cholate). If one looks at human bile fluid under a polarizing microscope, different phases can be observed (Olszewski *et al.*, 1973). A typical sample of fresh lithogenic specimens was found to have birefringent particles of two types. Small solid particles (cholesterol monohydrate crystals) in the form of parallelepipeds were identified as one of the birefringent particles. The second type of birefringent particle appears as batonets with diameters of about 0.5 μm; these textures are distinctively liquid crystalline. The likely structure of the liquid crystalline state is lamellar.

The role of liquid crystals in human bile fluid is somewhat speculative. However, it seems reasonable to suggest that the cholesterol concentration in the liquid crystalline phase is considerably higher than in the surrounding isotropic phase. Therefore, one role of the liquid crystal might be that it inhibits the precipitation of cholesterol. Further, one might look upon the liquid crystalline phase as a precursor to the formation of gallstones.

The liquid crystalline phase in bile fluid does not have a long stability. Holding the system at 37°C for several weeks showed a disappearance of the liquid crystalline phase with the formation of cholesterol monohydrate crystals.

Human bile fluid is a dynamic system involving several components. As pointed out in the section on hardening of the arteries, lecithin and water can form a system that solubilizes cholesterol. It has also been shown that bile salts in water solubilizes cholesterol (Lecuyer and Dervichian, 1969). Therefore, it is logical to suggest that injections of lecithin or bile salts into human bile fluids should retard the precipitation of cholesterol or bring the crystals of cholesterol into "solution." Other chemicals will also solubilize cholesterol in aqueous media. If these observations are further refined, gallstone formation could be treated chemically.

Is there a relationship between the presence of cholesterol in human bile fluids and its presence in the circulatory system? Is the presence of cholesterol and cholesteryl esters in bile fluid a precursor of their appearance in the blood and the harbinger of the appearance of hardening of the arteries or vice versa? These questions have not yet been answered, but the answers may be within our grasp.

C. Sickle Cell Anemia

Mutation of the gene that controls hemoglobin synthesis causes sickled cell anemia. Sickled cell anemia is a recessive familial disease inherited according to Mendel's law. Persons who have the disease have red blood cells which are abnormal (hemoglobin S or HbS) and which differ chemically from ordinary blood by one of 146 amino acids on each of the β-chains. Glutamic acid in the normal blood is replaced by valine on each of the β-chains. In a deoxygenating atmosphere, disc-shaped cells in sickled blood become crescent, or sickle-shaped, clogging small blood vessels and hampering oxygen transport. Under an oxygenating atmosphere, the disc-shaped cell cannot be regenerated if the cell is sickled; an ordinary cell that has been deoxygenated will, on oxygenation, return to the disc-shaped geometry.

The sickled cell in an aqueous environment is liquid crystalline in nature. Several different chemical methods have been used to treat sickled cell anemia. These include urea and sodium cyanate. The mechanism by which these chemicals function is not known.

D. Diseases Involving Liquid Crystal Accumulation

Many diseases accumulate liquid crystals. Small (1977) lists many of these, including (1) Krabbe's Disease, (2) Fabray's Disease, (3) Tangier's Disease, (4) Wolman's Disease, and (5) Gaucher's Disease.

All diseases that accumulate liquid crystals involve a specific block in the catabolism of particular lipid molecules. When the metabolic process is functioning effectively the lipid composition remains essentially at a fixed concentration and only in the case of disease is the metabolism in an imbalance and a massive build-up of the lipid in the organ results.

E. Ageing

The ageing process has always been uppermost in the minds of all who reach middle age. It is not known how the ageing process takes place, but the liquid crystalline state of matter must certainly be considered.

If one starts with an amphiphilic molecule in the crystalline state and adds water, a lamellar (bilayer) structure can be generated. Addition of more water can convert the liquid crystalline state from the lamellar to the hexagonally structured state. Still more water can convert the hexagonal structure to micelles and finally with more water to the true solution. Moving from the lamellar structure to the hexagonal structure, or going in the reverse direction, the structure can be altered as a function of water composition. This system is also temperature dependent and both parameters may function simultaneously or independently.

Ageing may, therefore, be a matter of total water concentration in the cell. As one goes from the hexagonal structure to the lamellar structure, the water is removed, and then to the crystal structure as more water is removed. As suggested, as ageing progresses, the cell membrane drifts toward crystallinity. If this is a correct though simplified explanation, then ageing could be delayed or modified, chemically by hydration and dehydration.

F. Cancer and Liquid Crystals

A word of caution is necessary in any attempt to discuss the subject of cancer. Ambrose (1976) has strongly indicated a relationship of abnormal cellular growth to cancer and has suggested that understanding the phenomena lies in the fact that liquid crystals can supply some information about the process. Ambrose points out that, in tissue culture of normal epithelial cells, the cells have continuous contact within the cell borders (membranes), but in cancerous cells there is a reduced cell to cell adhesiveness. Therefore, one of the changes from normal cells toward malignant cells is the appearance of gaps between cells. Furthermore, he has suggested other characteristics to be considered, such as a change of phase at the cell membrane surface resulting in a change of shape, a change in the filamentous fibrous proteins associated with cellular organelles and cellular mobility. These changes are characteristic of liquid crystalline properties. Therefore, any disturbance in the liquid crystalline state of normal cells could bring about a change leading to abnormal growth changes and hence malignancy.

IV. REMARKS

An all-purpose temperature sensor has been one goal of medical researchers. The liquid crystal meets this need to a great degree. One has available with liquid crystals a continuous monitoring of the skin temperature over an extended area. This provides the physician with a detailed and more easily interpreted indication of the circulatory patterns than can be obtained by point measurements with thermocouples and thermistors.

Liquid crystals can serve as detectors of contaminants in an atmosphere. Many chemicals that diffuse into a cholesteric–nematic liquid crystalline film affect the pitch of the helix and thus the color of the reflected light. An impurity can be observed if present in parts per million. The major study of this property should focus on selectivity. Selectivity problems raise questions such as whether methanol can be detected in the presence of ethanol, hydrogen chloride in the presence of hydrogen bromide, or carbon monoxide in the presence of oxides of nitrogen.

Only in recent years have liquid crystals been detected in diseases. Once our knowledge of the role of liquid crystals in diseases is expanded we should be able to treat certain diseases by simple chemical methods. Diseases evidently alter the liquid crystalline structures in cells, tissues, and organelles. Many biochemical studies related to disease have ignored the fact that the reaction is taking place in a liquid crystalline environment and not only in water. Once we accept the functioning of liquid crystals as solvents for biological reactions we can make some important steps forward in understanding certain diseases.

REFERENCES

Ambrose, A. J. (1976). Liquid crystals and cancer. *Adv. Chem.* **152,** 142ff.

Elser, W., and Ennulat, R. D. (1976). Selective reflection of cholesteric liquid crystals. *Adv. Liq. Cryst.* **2,** 73.

Hata, Y., and Insull, W. (1973). Significance of cholesteric esters as liquid crystals in human artherosclerosis. *Jpn. Cir. J.* **37,** 269.

Lecuyer, H., and Dervichian, D. G. (1969). Structure of aqueous mixtures of lecithin and cholesterol. *J. Mol. Biol.* **45,** 39.

Olszewski, M. F., Holzbach, R. T., Saupe, A., and Brown, G. H. (1973). *Nature (London)* **242,** 336.

Small, D. M. (1977). Liquid crystals in living and dying systems. *Colloid Interface Sci.* [*Proc. Int. Conf.*], *50th, 1976* Vol. 1, p. 615ff.

Chapter 13
Summary

I. OVERVIEW

The liquid crystalline state in living systems has only recently begun to be researched in some depth, mostly owing to the technological developments in liquid crystals. What we tried to do in the various topics covered was to point out that the physical state of matter involved in life processes is to a great extent liquid crystalline. In doing so, analogies were made between biological structures and the behavior of living cells to liquid crystalline structures and their behavior. In order to do this, the physical, chemical, structural, and optical properties of liquid crystals were discussed in some detail (Chapters 2, 3, 4) to emphasize their special properties.

We then examined the structure of the principal molecules we associate with life systems. The important point here is how these molecules became ordered into membranes, and we conjectured how such membranes could have given rise to the primordial cell (Chapters 5 and 6). In our discussion of the cell (Chapter 7), some thought was given to how the primitive prokaryotic cell gave rise to various cellular organelles found in the more highly evolved eukaryotic cell. As a result, we described the structure of these organelles, which are fundamental to the cell function. Emphasis was placed on the cell membrane and the membranes of these various cellular organelles, such as the nucleus, mitochondria, Golgi apparatus, endoplasmic reticulum, and other cytoplasmic structures.

We then considered such differentiated organelles of the cell that are specialized for photoreception. These were the chloroplasts for photosynthesis (Chapter 8) and the retinal photoreceptors of the eye, the rods and cones for visual excitation (Chapter 9). These photoreceptors are all structured of double membranes that form lamellae (Figs. 8.2, 8.3, and 9.2–9.4). The photoreceptors, in invertebrate animals, arthropods, and molluscs—the rhabdomeres—comprise membranes in microtubular structures hexagonally packed (Figs. 9.8–9.11). Other receptor–effector structures such as muscle and nerve, which are part of the neuromuscular systems of animals, were then examined (Chapter 10). The nerve with its myelin sheath (Figs. 10.4–10.7) and the striated muscle protein structures of thick myosin filaments and thin actin filaments (Figs. 10.2 and 10.3) were described. It was indicated that the lattice of the thick filaments, myosin, is a smectic liquid crystalline structure.

What all of these structures, which are involved in the energenetics of the cell, reveal within the resolution of the electron microscope, is that they consist of a highly ordered lamellae, membranes, that show a lamellar, cubic, or hexagonal packing, and we emphasized that such structures are similar to structures formed by liquid crystals. This striking architecture at the molecular level must be related to their function and have a common physical–chemical origin for all life systems.

II. CHOLESTERIC–NEMATIC LIQUID CRYSTALS

The cholesteric–nematic liquid crystalline state is one in which the orientation of the molecules is in planes parallel with the planes twisting to form a helicoid, a periodic structure.

Molecules of interest in biological structures such as the polysaccharide polymer chitin, polynucleotides, polypeptides, and proteins, as well as DNA and transfer RNA exhibit cholesteric–nematic structures.

The cuticle, exoskeleton of arthropods, invertebrate connective tissue, plane cell walls, the pellicle of protozoan flagellates, and vertebrate bone have a helical periodic structure. The corneal lens of arthropod eyes important for vision of these animals, has a helical spiral. Such helical structures depending on their pitch exhibit interference colors. The corneal lens structure (Figs. 9.14–9.17) could function then as a polarizer of light for navigation.

The helical structure is probably a relatively common architecture found from molecules to biological structures. These structures indicate the importance of understanding the liquid crystalline state in interpreting how these biological structures are formed and function.

III. MEMBRANES

The integrity of the cell depends on its cell membrane and hence is of primary importance to life. The cell membrane separates the external environment from the internal environment. In the process, the cell membrane must allow for the differential diffusion of ions, molecules, and the exchange of gases, and it has selective properties. Therefore the cell membrane plays an active role in chemical transport, energy transduction, and information transfer to and from the cell.

The cell membrane structure of lipids and proteins was discussed in detail in Chapter 11. Lipids in water form liquid crystal structures like that of myelin. Membrane function may be closely associated with these lipids. The membrane lipids can undergo phase transitions depending on temperature. Lipids in water at low temperatures are in a solid state and at higher temperatures they are in a liquid crystalline smectic state. The temperature of the phase transition from the solid state to the liquid crystalline state depends on the kind of lipids as well as the fatty acid composition of these membranes. In general, the higher the degree of unsaturation of fatty acid, the lower the temperature for phase transition, because a cell membrane containing phospholipids with little unsaturation is less fluid than one with much unsaturation. The control of fluidity of the lipid components may be related to the diffusion of molecules and ions passing in and out of the cell membrane. Those organisms that alter their internal body temperature to correspond to their external environment may do this by altering the degree of saturation of the phospholipids in their cell membrane. Correlation between the phase transition temperature and the fatty acid composition has been found in biological membranes where changes of the physiological activities are observed at the phase transition temperatures. The membrane lipids play an important role in the functioning of the photosynthetic reactions. In the chloroplasts of most algae and higher plants there is a relatively high concentration of unsaturated lipids. Also, in the chloroplast lamellae, like that of the cell membrane, a phase transition of the lipids with temperature occurs.

Membranes are not static fixed structures, for their molecules have a degree of mobility. Labeling experiments show that the membrane components do not exchange rapidly from one side of the bilayer to the other, but most other motions are available. The dynamic nature of the membrane is observed, for example, in that the fastest molecular motion is about the C–C bonds in the lipid hydrocarbon. The lipids can exhibit translational motion and can undergo fast lateral diffusion within the plane of the membrane as well (Fig. 11.1). Proteins also have mobility but are slower than the lipids.

To arrive at a structural molecular model for the cell membrane is not a simple task. We did, however, schematize the various cell membrane molecular models that seem to best fit the experimental data at our present level of understanding (Fig. 11.2). The important structural basis is that cell membranes are lipid bilayers to which specific proteins become a part, due to their structure membranes exhibit properties that we associate with liquid crystalline states.

IV. ANALYTICAL TOOL

Liquid crystals can be used other than as model systems for interpreting the molecular structure and function of various cellular organelles, tissues, and organs. A direct application to medicine is in measuring body temperature, in locating areas of the body which have temperatures different than their surroundings. Hence, liquid crystals can be a powerful analytical tool. In addition to measuring temperature changes, they can be used to determine pressures, as well as electrical and magnetic fields. Thus, they function as sensors and are implicated in the sensory systems for vision, smell, and touch.

V. CONCLUSION

The conditions necessary for the molecule to organize and form membranes and cellular structures cannot be found in amorphous isotropic liquids. A liquid crystalline state, however, would provide the structural organization and the mobility required by the living state.

". . . a liquid crystal in a cell, through its own structure, becomes a proto-organ for mechanical, chemical or electrical activity, and when associated in specialized cells (with others) in higher animals gives rise to true organs, such as muscle and nerve." Second, and perhaps more fundamentally, ". . . the oriented molecules in liquid crystals furnish an ideal medium for catalytic action, particularly of the complex type needed to account for growth and reproduction" Bernal (1933, p. 1082).

We have hardly covered all the aspects of liquid crystalline states and their importance to life processes. There is no doubt that liquid crystalline states associated with plant and animal structures hold great promise for the elucidation of the developmental aspects of growth and differentiation so important to understanding life systems.

Further vigorous exploration of the liquid crystalline state of biological structures is required. We hope this will lead to a better understanding of many of the complexities that we associate with life itself.

REFERENCE

Bernal, J. D. (1933). Liquid crystals and anisotropic melts. *Trans. Faraday Soc.* **29**, 1082.

General References

Blumstein, A., ed. (1978a). "Liquid Crystalline Order in Polymers." Academic Press, New York.

Blumstein, A., ed. (1978b). "Mesomorphic Order in Polymers and Polymerization in Liquid Crystalline Media," ACS Symp. Ser. No. 74. Amer. Chem. Soc., Washington, D.C.

Brown, G. H. (1967). Liquid crystals. *Chemistry* **40**, 10.

Brown, G. H. (1969). Liquid crystals and some of their applications in chemistry. *Anal. Chem.* **41**, 26A.

Brown, G. H. (1972). Liquid crystals and their roles in inanimate and animate systems. *Am. Sci.* **60**, 64.

Brown, G. H., ed. (1975). "Advances in Liquid Crystals," Vol. 1. Academic Press, New York.

Brown, G. H., ed. (1976). "Advances in Liquid Crystals," Vol. 2. Academic Press, New York.

Brown, G. H., ed. (1978). "Advances in Liquid Crystals," Vol. 3. Academic Press, New York.

Brown, G. H., and Doane, J. W. (1974). Liquid crystals and some of their applications. *Appl. Phys.* **4**, 1.

Brown, G. H., and Shaw, W. G. (1957). The mesomorphic state. *Chem. Rev.* **57**, 1049.

Brown, G. H., Doane, J. W., and Neff, V. D. (1971). "A Review of the Structure and Physical Properties of Liquid Crystals." CRC Press, Cleveland, Ohio.

Chandrasekhar, S. (1977). "Liquid Crystals." Cambridge Univ. Press, London and New York.

de Gennes, P. G. (1974). "The Physics of Liquid Crystals." Oxford Univ. Press, London and New York.

Demus, D., and Richter, L. (1978). "Texture of Liquid Crystals." Verlag Chemie, Weinheim.

Fergason, J. L., and Brown, G. H. (1968). Liquid crystals and living systems. *J. Am. Oil Chem. Soc.* **45**, 120–127.

Frey-Wyssling, A. (1953). "Submicroscopic Morphology of Protoplasm." Elsevier, Amsterdam.

Frey-Wyssling, A. (1957). "Macromolecules in Cell Structures." Harvard Univ. Press, Cambridge, Massachusetts.

Gray, G. W. (1962). "Molecular Structure and the Properties of Liquid Crystals." Academic Press, New York.

Gray, G. W., and Winsor, P. A., eds. (1974). "Liquid Crystals and Plastic Crystals," Vols. 1 and 2. Wiley, New York.

Hartshorne, N. H. (1974). "The Microscopy of Liquid Crystals." Microscope Publications, Ltd., Chicago, Illinois.

Meier, G., Sackmann, E., and Grabmaier, J. G. (1975). "Applications of Liquid Crystals." Springer-Verlag, Berlin and New York.

Nagoette, J. (1936). "Morphologies des Gels Lipoides." Hermann & Cie, Paris.

Needham, J. (1968). "Order and Life." MIT Press, Cambridge, Massachusetts.

Porter, K. R., and Bonneville, M. A. (1964). "Fine Structures of Cells and Tissues," 2nd ed. Lea & Febiger, Philadelphia.

Priestley, E. B., Wojtowicz, P. J., and Sheng, P., eds. (1975). "Introduction to Liquid Crystals." Plenum, New York.

Prince, L. M., and Sears, D. F., eds. (1973). "Biological Horizons in Surface Science." Academic Press, New York.

Wolken, J. J. (1971). "Invertebrate Photoreceptors: A Comparative Analysis." Academic Press, New York.

Wolken, J. J. (1975). "Photoprocesses, Photoreceptors and Evolution." Academic Press, New York.

Index